GENDER MOSAIC

GENDER MOSAIC

Beyond the Myth of the Male and Female Brain

DAPHNA JOEL, PhD
and LUBA VIKHANSKI

Little, Brown Spark

New York Boston London

Little, Brown Spark
Hachette Book Group
1290 Avenue of the Americas, New York, NY 10104
littlebrownspark.com

First Edition: September 2019

Little, Brown Spark is an imprint of Little, Brown and Company, a division of Hachette Book Group, Inc. The Little, Brown Spark name and logo are trademarks of Hachette Book Group, Inc.

The publisher is not responsible for websites (or their content) that are not owned by the publisher.

The Hachette Speakers Bureau provides a wide range of authors for speaking events. To find out more, go to hachettespeakersbureau.com or call (866) 376-6591.

10 9 8 7 6 5 4 3 2 1

ISBN 978-0-316-53461-1
LCCN 2019933948

LSC-C

Printed in the United States of America

To our mothers

Contents

Contents

I

Sex and the Brain

1.

My Awakening

One summer morning over ten years ago I was home with my three kids when I heard the hissing of a burst water pipe outside my apartment building. Holding my youngest, then still a baby, in my arms, I rushed to the yard, bent the end of the gushing rubber pipe from an automatic garden-watering system to stop the water running, and asked my two older kids to call a friend who lived next door. When my friend arrived, I assumed he'd take care of the situation. But he just stood there, obviously at a loss. Only then did it dawn on me that he was just as clueless about plumbing as I was. I asked him to hold the folded pipe—and the baby—and went looking for the main faucet so I could shut off the water.

It had taken the bewildered look on my friend's face for me to become aware of my own bias. I must admit I was embarrassed. I had always believed in equality between the sexes, and I thought I was running my life accordingly. Yet here I was, expecting a man to handle a technical emergency.

Just around that time, I received an excellent opportunity to explore in depth my own and other people's gender biases: a colleague asked me to take over a course on the psychology of gender she had been teaching at Tel Aviv University. To prepare, I spent a year reading books and scientific articles on the development of women and men from the moment of conception. As a neuroscientist, I was most interested in the relationship between sex and the brain.

I soon learned that many scientists, just like many others, believe that the brains of females and males differ in profound and important ways, and that this is the reason for fundamental differences between women and men in almost every domain, from cognitive and emotional abilities, through interests and preferences, to behaviors. Self-help books that try to teach us how to cope and communicate with the other sex tend to take this belief as a given.

According to a popular version of the story, the female brain has a large communication center and a large emotion center, and is hardwired for empathy. The male brain, on the other hand, has a large sex center and a large aggression center, and is hardwired for building systems.

This story seems to provide us with a neat biological explanation for much of what we encounter in day-to-day life. It explains why women are more sensitive and emotional, whereas men are more aggressive and sexual; why most teachers are women, and most engineers, men.

"It's the hormones, stupid," we are told. In the womb, the story goes, a huge surge of testosterone, secreted by the testes of the male fetus, changes his brain from the default female to male. So girls are born with a female brain and boys with a male brain. The details of the rest of the story vary among different authors, but they all explain why women and men behave in the ways depicted by popular gender stereotypes. Girls are nice and empathic, and boys are active and aggressive—because this or that region in the brains of girls is smaller or larger than in boys, or because they have more or less of this or that hormone.

There are no surprises. No matter what the findings are, they are never interpreted in a way that would run counter to prevalent gender stereotypes. For example, the amygdala, a brain region central to emotions, was for many years considered to be, on average, larger in males than in females, yet no one ever claimed that based on the size of their amygdalae, men were, by nature, the more emotional sex. (Recent statistical analyses have shown that, in fact, there is no difference between the sexes in the average size of this brain region.)[1]

The concept of a male brain and a female brain fits well the popular view that men and women come from different planets, but does it fit the scientific evidence? My own attempt to answer this question began with a startling study I encountered about a decade ago while preparing to teach the psychology of gender course.

Did you know that thirty minutes of stress is enough to change the "sex" of some brain regions from male to female, or vice versa? I hadn't known this either. Reading about this study led me to several years of extensive research, which completely transformed the way I thought about sex, gender, and the brain.

After analyzing hundreds of brain scans, I realized that sex differences in the brain do not consistently add up in individuals to create "male" and "female" brains. Note that I'm not saying there are no differences between the brains of females and males; on the contrary, my team has documented many such differences, as have numerous other scientists. What I'm arguing is that these differences mix together in each individual brain to create a unique mosaic of features, some of which are more common in females and others are more common in males. This notion goes hand in hand with what I'm sure many people already know: we are all patchworks of "feminine" and "masculine" traits. But it goes further; it suggests that there is no such thing as a "male" or "female" brain—or a "male" or "female" nature.

Before I describe how I arrived at the brain mosaic idea and what its implications might be, let me share with you a few intriguing facts about the brains of females and males, and how the perception of these facts has changed over the past few centuries.

2.

A History of
Twisted Facts

When egalitarian ideas started circulating prominently in Europe in the seventeenth and eighteenth centuries, men faced an embarrassing dilemma. The new principles implied that all humans, women and men, were by nature equal. This notion threatened the existing social order, in which women played subordinate roles. The fear was that equality would undermine the very foundations of society—most important, that given equal status, women would stop serving men.

Molière satirized these fears in his 1672 comedy *Les Femmes savantes,* in which the husband rails against his wife and other science-minded women, who neglect their domestic duties: "They want to write and become authors. No science is too deep for them…They know the motions of the moon, the pole star, Venus, Saturn and Mars…and my food, which I need, is neglected."[1]

Science was called upon to resolve the political debate over the role of women in an egalitarian social order. In *The Mind Has No Sex?*, Londa Schiebinger, of Stanford University, writes that the mission was to show that it was nature, not men, that was responsible for gender inequality. Schiebinger traces how the scientific study of female and male anatomy, including the brain, turned political. Without abandoning the axiom of equality, she argues, the medical and scientific communities became preoccupied with differences between the sexes. "Women were not to be viewed merely as *inferior to* men but as fundamentally *different from,* and thus *incomparable to,* men," she writes.[2]

Sexual differences between females and males are all too obvious, but do they extend to the entire body and the brain? A great deal was at stake: answering this question in the affirmative could help justify the different social standing of women and men; a negative answer would suggest that women had been unjustly oppressed for centuries, and that major social changes were needed. A great many philosophers and other thinkers—virtually all of them male—tended to define the scope of the differences between the sexes in the broadest possible terms. Schiebinger quotes one eighteenth-century French physician as saying that "the essence of sex is not confined to a single organ but extends, through more or less perceptible nuances, into every part."[3]

Science became a legitimate arbitration arena for such disputes. Unlike religion, which had carried the burden of justifying women's inferiority up to the scientific revolution, science was thought to be impartial, and therefore to provide objective evidence in the arguments over women's abilities. "Perhaps the knife of the anatomist could find and define sexual difference once and for all," Schiebinger writes. "Perhaps sexual differences—even in the mind—could be weighed and measured."[4]

Indeed, wrote Stephanie Shields, of Pennsylvania State University, weighing and measuring the skull and, later, the brain—by then established as the seat of the mind—became of paramount importance.[5] In ancient Greece, Galen had deemed the testicles the most noble part of the body—which made perfect sense because they were found only in the "superior" sex. But in the seventeenth century, it was the brain that came to be viewed as the most noble and divine organ: holder of all the senses, intelligence, and wisdom. It was hence essential that men be found to have superior brains.

Initially, this seemed like an easy task. The skull— thought to provide a reliable indication of brain size— was found to be, on average, smaller in women than in men. What could better explain women's inferiority (well, except for the absence of testicles)?

But it was too early to celebrate. After all, quite a few animals have larger skulls than we do. Sperm whales, for one, have skulls that are, by far, larger than those of humans. Scientists who were keen on proving the superiority of men over women—but surely not the superiority of whales over men—searched for a way around this inconvenient fact. They suggested that perhaps it wasn't the size of the skull but the ratio of skull size to body size that mattered.

Yet calculating the ratio failed to produce the desired results. Worse still, a number of scientists actually found that relative to total body weight, women's skulls were larger than men's. These scientists did not for a moment conclude that the relatively larger skulls of women meant greater intelligence. Undeterred in their zeal to produce "scientific" evidence of male superiority, some scientists managed to interpret their findings as a sign of women's *lesser* intelligence. Women, they said, resembled children, whose skulls were also large relative to body size, which meant that women were less developed, and consequently less intellectually competent than men.

Looking back at the history of brain research, I'm impressed by the creativity that went into twisting scientific facts to serve a social or political agenda. When scientists didn't like what they found, they often either changed the interpretation or simply abandoned the method that led to the undesired result, looking instead for a "better" measure. According to Shields, reams of paper were dedicated to arguing over "appropriate" measures for skull size in women and men. Should it be the ratio of skull weight to body weight? Perhaps it was a matter of bone density in the skull compared to the rest of the skeleton? The issue proved impossible to resolve: applying some measures, the results "favored" men; with others, they "favored" women.

The idea that larger is better continued to be popular when scientists discovered that not only the skull but also the brain was larger, on average, in men than in women. The eminent nineteenth-century neuroscientist Paul Broca was among the more diplomatic, but he still expressed this view in no uncertain terms. "We might ask if the small size of the female brain depends exclusively on the small size of her body," Broca wrote in a scientific journal in 1861. "But we must not forget that women are, on the average, a little less intelligent than men, a difference that we should not exaggerate but which is nonetheless real."[6] The prominent evolutionary biologist George Romanes was blunter. The smaller size of women's brains was responsible for female mental inadequacy, he wrote in 1887, which "displays itself most conspicuously in a comparative absence of originality, and this more especially in the higher levels of intellectual work."[7] Theodor Bischoff, a distinguished nineteenth-century biologist, even went as far as to claim that women, because of their small brains, did not have the intellectual skills necessary for academic studies, and that too much education might hinder the development of reproductive organs in adolescent girls.[8]

These older versions of the belief in female and male brains being fundamentally different sound absurd to us now. Today, when women outnumber men at so many levels of academic study, it seems ridiculous that scientists could have believed women to be incapable of attending university because of the size of their brains. Don't get me wrong; women's brains are, on average, still smaller than men's. What has changed is not the size of the brain, but rather the social norms that once prohibited or discouraged women from studying.

While the focus on brain size took on a life of its own, the search for scientific findings in support of men's superiority over women had, in the meantime, moved into a new arena. In the wake of the discovery in the nineteenth century that different brain regions perform different functions, scientists began comparing these regions in women and men. Unsurprisingly, here too they found anatomical support for the superior intelligence of men.[9]

Much of the attention was directed to the cerebral cortex, as this outer part of the brain is responsible for voluntary action, perception, cognition, language, and thought. It is made of so-called gray matter, which holds the bodies of billions of nerve cells—the neurons. Underneath the cortex there is a layer of white matter, which mainly contains the fibers connecting the neurons. The cortex has traditionally been subdivided into four major lobes, named after the skull bones protecting them.

When the role of the frontal lobes in cognitive function was established, many neuroscientists were quick to point out that these lobes were larger and more developed in men than in women. Then some neuroscientists suggested that the seat of the intellect resided in the parietal lobes, those at the top of the brain, rather than in the frontal lobes. And once the importance shifted to the parietal lobes, certain scholars promptly revised the interpretation of anatomical findings to match the accepted view of male superiority.[10] In 1895, for instance, American psychologist George Thomas White Patrick wrote in *Popular Science Monthly* that "the frontal region is not, as has been supposed, smaller in woman, but rather larger relatively....But...a preponderance of the frontal region does not imply intellectual superiority...the parietal region is really the more important."[11]

More than a hundred years have passed since these words were written. In that time, neuroscientists have continued to find differences between the brains of males and females, in animals and humans. I'll discuss them again in the next chapter, but here are a few examples. Most of the cortex is, on average, thinner in men than in women; men have on average a lower proportion of gray matter and a higher proportion of white matter. In addition, men have larger ventricles—big, fluid-filled cavities in the center of the brain (these are the large solid-color spaces you see on medical scans). Readers who were happy to learn that men have larger brains than women may be less happy to read about men's larger ventricles.

If you believe, as did scientists in the nineteenth century, that the size of the brain matters, then it is indeed awkward to learn that your bigger brain comes packaged together with bigger—what should I call them?—empty spaces. But the message I want to convey is that both sexes have nothing to worry about. Men do just fine with their larger ventricles; women do just fine with their smaller brains.

What *is* worrying is that sex differences are still being used to justify gender inequality. No one today would dare use biological comparisons between races or social classes to justify racism or the economic status of the poor—as was done up to the twentieth century—but sex differences in the brain are still being pulled out to validate women's inferior status. Here's how Schiebinger puts it: "The alleged defect in women's minds has changed over time: in the late eighteenth century, the female cranial cavity was supposed to be too small to hold powerful brains; in the late nineteenth century, the exercise of women's brains was said to shrivel their ovaries. In our own [that is, the twentieth] century, peculiarities in the right hemisphere supposedly make women unable to visualize spatial relations."[12] In the twenty-first century, the search for the "essential" difference between the brains of women and men continues, resonating all too often with historic myths about differences between the sexes.

3.

As the Differences
Pile Up

A few years ago, I took part in a scientific discussion, "SeXX and SeXY: A Dialogue on the Question of the Female Brain and the Male Brain," within the framework of a neuroscience symposium at Stanford University.[1] My co-discussant, Louann Brizendine, argued that women are women because they have a female brain, and men are men because they have a male brain—a view she has also expressed in her bestselling books. I, for my part, presented my own view: that humans and their brains are composed of unique mosaics of "female" and "male" features. After the debate, I overheard someone tell one of the organizers: "The problem is that Louann has a female brain, and

Daphna has a male brain." This person was probably implying that Brizendine and I weren't well matched as public speakers because of our different debating styles.

But here's the irony. This remark undermined the view held by Brizendine and many others—that the male brain is a product of exposure to high levels of testosterone in the womb and later in life, whereas the female brain develops as a default, in the presence of low testosterone levels in the womb, and is later further shaped by "female" hormones. If that were the case, how could I, born a typically developed female, who later on had been exposed to high levels of "female" hormones in the course of three pregnancies and a total of about three years of breast-feeding, possibly have a male brain?

Irony apart, the belief that "men's brains are like *this,* and women's brains are like *that,*" is still immensely popular, among scientists and the general public alike. Today—as in previous centuries—the common belief is that differences between the brains of men and women lie at the root of the fundamental differences assumed to exist between the sexes. It is little wonder, then, that this area of research is so intense. A review of scientific literature published in *Neuroscience & Biobehavioral Reviews* in 2014 yielded some 5,600 studies in which the volume and density of various brain regions in men and women had been compared in the preceding quarter of a century.[2]

By now scientists have reported on hundreds of sex differences in the brain. Women and men differ in the size of the entire brain and in the size of specific brain regions. (Many of the latter differences disappear when overall brain size is taken into account; others are reduced or even reversed—that is, a region may be smaller, on average, in women, but relatively larger compared to their brain size.)[3] With the advance of technologies enabling scientists to peer into the brain with increasing depth and detail, sex differences have also been found in several systems of chemical messengers called neurotransmitters. Moreover, differences have been found in the microanatomy of the brain—the structure of neurons and the density of receptors, the molecules to which the neurotransmitters bind.

Note that these are all average differences—they emerge when women and men are compared as two groups, but not necessarily when they are compared individually. We may, for example, find that a particular brain region is, on average, larger in men taken as a group than in women as a group. But if we make individual comparisons, we'll discover a great deal of overlap—namely, that this region is the same size in many women and men; furthermore, in some women it will be large, whereas in some men it will be small. That's true for most known sex differences in the structure of the human brain. The average differences are small, and there is a great deal of overlap between the sexes.[4]

There are scientists, however, who argue that although average sex differences in brain structure are small, they may underlie greater differences in brain function—in other words, that the brains of women and men, even if similar in structure, might *work* differently. That's the rationale for studies aimed at discovering sex differences in patterns of brain activity during the performance of various mental tasks.

But in reality, in most tasks, the patterns of brain activation are similar in women and men; lots of studies have looked for such differences but failed to find any. When studies do find a difference, it is typically present in the functioning of only some of the brain regions involved in performing a certain task, while others are similarly activated in the two sexes. The problem is that the sameness, for the most part, goes unreported, whereas the differences get plenty of play in both the scientific and the popular press.

That was just what happened with one popular theory: that when processing language, women tend to use *both* cerebral hemispheres to a greater extent than men. For instance, in a study published in 1995 in *Nature,* Yale University researchers used a method called functional magnetic resonance imaging to scan brain activity in nineteen women and nineteen men who were asked to perform three types of language-related tasks.[5] In their scientific paper, the researchers devoted little attention to the two tasks—letter recognition and grouping words by meaning—in which they found no difference between the sexes. Instead, they reported in great detail on the third task—rhyming—for which they did find a sex difference: when performing this task, men activated a number of regions on the left side of their brain; women activated these same regions on both sides of the brain. The brain scans for rhyming, which lit up in different patterns in the two sexes, were included in the paper; the scans for the other two tasks were not.

The study got plenty of media attention—it fit in nicely with the stereotype of the sexes being different to the point of applying their brains to a given task in different fashions. Some of you may even recall those newspaper headlines and stories on TV. Here's a fairly dramatic one from the *New York Times:* "Men and Women Use Brain Differently, Study Discovers."

Then a number of other studies showed no consistent difference in brain activity between women and men performing various language-related tasks. (How such disparate results come about is an interesting issue in itself; I'll discuss it in chapter 8.) In an attempt to resolve the controversy, scientists from the University Medical Center Utrecht pulled together the results of twenty-six studies on this topic using a statistical method called "meta-analysis." Their conclusion, published in 2008 in a journal called *Brain Research,* was that no difference in language processing could be proven to exist between the sexes.[6] Do you recall any major media stories about these findings? I don't either.

Such selective reporting—that is, playing up the differences and ignoring the similarities—is usually unintentional; it occurs because writing about the differences is much more interesting than reporting that no differences have been found. But it ends up creating the impression that the differences between the sexes are much greater than they truly are.

Yet some scientists argue that even if each of the differences is small and even if there are few of them, together they add up to a large difference between women and men. In fact, this is what many people believe, even if only implicitly. Therefore, as more and more sex differences in brain structure and function were discovered in the past few decades, the belief in the existence of male and female brains has kept getting stronger—because everyone was taking for granted that these differences were adding one to another to create two types of brains, female and male. But do the differences really add up in this manner?

In this book, I will argue that they don't. Certain brain features do differ, on average, between women and men, but as a rule, features that are more common in women don't consistently add up in women's brains; nor do features more common in men consistently add up in men's brains. Instead, these differences *mix* up, so that human brains—as well as human psychological

21

characteristics and behaviors—are mosaics of features, some more common in women, others more common in men. This conclusion has nothing to do with the thorny issue of why some features are more common in one sex than in the other.

4.

Nature vs. Nurture

I n one of the episodes of the BBC's motor vehicle television show *Top Gear* I happened to watch, the presenter joined a couple in their forties in road-testing their sports sedan. Throughout the ride, the wife, delicate and feminine, elaborated on the car's power transmission, engine capacity, and various other technical parameters. As the presenter listened to her superbly informed commentary, his mien changed gradually—from one of utter disbelief to that of a smug showman who'd managed to serve up his audience a talking horse.

Indeed, men are, on average, more knowledgeable about cars than women, and they are better than women at telling one make of car from another. If we were to identify the neural network responsible for car expertise, we might find differences between men and women. But if so, would this reflect an inborn difference in this network between men and women, or an average lifelong difference in interest in cars?

This sort of question arises in connection with all the differences between men's and women's brains. Are they determined by nature (that is, by innate and preprogrammed biological factors) or by nurture (the environment in which we are raised and in which we live)? This is the "nature versus nurture" dilemma.

When it comes to the brain, we tend to reserve for nature a special place of honor, much more so than when considering other parts of our anatomy. For instance, people who work outside in the sun have, on average, darker skin than those who work in an office, but we don't conclude that dark-skinned people prefer outdoor jobs, and that light-skinned people prefer office jobs— because we know that the skin gets darker with exposure to the sun. Yet we commonly assume that a difference in brain structure or function is the cause of differences in observed behaviors, seldom considering the possibility that it may be the other way around.

Interestingly, there are hardly any differences in brain structure and function between female and male newborn babies, except for a difference in total brain size—it is larger, on average, in male babies by about 6 percent.[1] This does not rule out the possibility that sex differences

24

observed later in life in the human brain are preprogrammed. Note, for example, that there is no difference
in the shape of breasts in human females and males before
adolescence. But the appearance of sex differences in the
brain only later in life does mean that they might very
well reflect, at least in part, gaps between the experiences
of women and men throughout their lives.

Our brain is not a fixed, hardwired machine. Rather,
it is highly malleable and changes throughout our
lifetime—this wonderful property is known as plasticity. Therefore, it's not just that the brain affects our
behavior—our behavior, in turn, affects the brain.

You may have heard about studies of London taxi
drivers[2] concluding that years of memorizing hundreds
of routes and street names lead to an increase in the
volume of the hippocampus—a neural structure critical
for spatial abilities. The ongoing need to navigate the
complex maze of the city's streets had forced the drivers'
brains to cope with the challenging spatial experiences.
Even much shorter-term experiences can lead to detect-
able changes in the brain. In one study, researchers at
the National Institute of Mental Health in the United
States asked a group of adults to spend a quarter of an
hour every day tapping their fingers one by one against
the thumb in a designated sequence. After just three
weeks of such practice, the portion of the cortex that was
activated during this movement increased.[3]

Because I don't want to leave you with the impression

that more is always better in the brain, let me mention yet another study, by researchers at Georgetown University Medical Center. They found that when children become proficient readers, some of the neural circuits activated during reading tend to become less active, as the children increase their reliance on other brain regions.[4]

Imagine, then, to what extent and in what complex ways the brain may be shaped by the different experiences of girls and boys, women and men. In our society, in which boys and girls are treated differently from the moment of birth, and in which different behaviors are expected of the two sexes, it's impossible to tell whether a difference between females and males in a brain feature, cognitive ability, or behavior is innate (preprogrammed) or results from experiences and external influences.

Baby girls, for instance, do better than baby boys, on average, on verbal tests. One might assume that an advantage emerging so early in life should reflect an innate sex difference—until one learns that talking to babies is one of the most important factors in the development of language skills, and parents tend to talk to baby girls more than to baby boys. How, then, can we tell whether the girls' superior verbal skills indeed stem from their sex or whether they are affected by the gendered care they receive?

In popular usage, the words "sex" and "gender" are often used interchangeably, but since the 1970s, researchers have been drawing a distinction between the two.

Sex refers to the biological features that go with having male or female genitals and *gender* to the social features. Typically included under sex is a genetic component—whether a person has a pair of X chromosomes, or an X and a Y—and a hormonal component, covering such hormones as testosterone, estrogen, and progesterone.

The concept of gender originally referred to traits considered appropriate for males and females—that is, masculinity for men and femininity for women—but in the past few decades, this concept has been widened, so that gender is also recognized as a social system that affects such aspects of our lives as access to power and relations with others. In modern Western societies, men, as a group, possess more of such valued resources as land and money, form a majority in legislative bodies, and enjoy a higher social status than women.

As I was learning more about this contemporary understanding of gender, it became obvious to me that, as feminist scholars have claimed, the division of traits between the two genders is anything but random. Qualities associated with men are characteristic of a dominant group: strength, willpower, need for achievement, competitiveness, aggressiveness. Qualities associated with women are those of a subordinate group: weakness, gentleness, kindness, sensitivity, warmth, empathy, nurturing.

A vivid way of showcasing these gender inequalities is through spotting them in advertising. In his book *Gender Advertisements,* published in 1976, sociologist

Erving Goffman reveals how ads reflect the dominance of men and the subordination of women. In the more than five hundred photo advertisements he analyzed, women, for example, much more often than men, are pictured bashfully bending a knee or canting their body and head, a posture that can be read as "an expression of ingratiation, submissiveness, and appeasement."[5] In ads showing social interactions, men more often instruct women than the other way around, and they tend to be placed higher than women in the image, symbolizing a higher social status. Goffman argues that these displays don't reveal the "true nature" of women and men. Rather, they reveal how culture defines "what our ultimate nature ought to be."[6] Things may have changed somewhat since Goffman's book first appeared, but a documentary film based on his work, *The Codes of Gender*, which came out in 2010, presents plenty of current examples in which ads show men in dominant poses and women in subordinate ones.

Gender affects not just our behavior but also various aspects of our physiology. For example, the average difference in muscle mass between women and men depends partly on sex and partly on gender, which has traditionally imposed on females and males different norms for physical labor and working out. Broadly, boys and men are expected to be strong and athletic, to lift weights or at least look as if they could. Girls and women are generally expected to be gentle not only in character but also in looks, so much so that some of them may worry about bulking up if they exercise a great deal.

Even some aspects of sex itself,[7] such as its hormonal component, are influenced by gender.[8] Blood levels of testosterone, for instance, commonly viewed as the biological "essence" of masculinity, are subject to multiple external factors, some of which are gendered. Thus, engaging in competition—still often expected of men but discouraged in women—has been shown to alter testosterone levels.[9]

And, as I found out in the course of my reading, gender affects the brain. For example, mothers and fathers are widely regarded as having different parenting styles. But are such differences preprogrammed in our biology, or dictated by the roles allotted to women and men in our society? One highly publicized study, reported in 2014 in the *Proceedings of the National Academy of Sciences, USA,* was carried out in the laboratory of Ruth Feldman, of Bar-Ilan University. She and her colleagues found that in heterosexual couples raising their first-born infant, patterns of brain activity differed somewhat between mothers and fathers.[10] Interestingly, in gay fathers acting as primary caregivers to their infants, there were similarities to activity patterns of *both* the heterosexual mothers and fathers. In other words, the way in which the brains of the men in the study responded to parenthood depended at least partly on parenting roles, which are commonly determined by gender.

So gender-related experiences can clearly affect the brain, but this doesn't mean that sex doesn't produce its effects as well. However, to prove that sex, and not gender, is responsible for a difference emerging later in life, we'd need to compare the brains of females and males who'd grown up and lived their entire lives in a gender-free society. Even if we could carry out such an incredible experiment, what would we do about its results?

If we discover a mutation that causes difficulties in acquiring reading skills, would we give up on teaching children with this mutation how to read? We'd more likely do just the opposite—use our knowledge to identify these children as early as possible and do our best to help them overcome their inborn disadvantage. Why, then, treat sex effects differently? If we were to discover that a certain innate brain feature predisposes people to violence, and that this feature is more common in boys than in girls, should this discovery increase our tolerance toward violence in men? Surely not. Rather, I believe we would provide children born with this feature, boys or girls, extra help to overcome their inborn violent tendencies. Knowing the biological basis of violence may allow us to develop better intervention methods for these children. But it would not provide them with a license to kill.

Unfortunately, the ongoing preoccupation with the origin of sex differences does not seem to be driven by a wish to improve human lives. Rather, just as in previous centuries, it is fueled—for the most part unconsciously—by the desire to justify the social inequalities between the sexes.

I hope to have convinced you that nature versus nurture is an unresolvable dilemma, and that it keeps many people busy for the wrong reason, namely, preserving the existing social order. But the real reason I see no point in dwelling on this issue is that it has no bearing on the central question being posed: Do brains come in two forms, female and male?

To answer this question, we need to determine whether the various sex differences in the brain—whatever their source—add up consistently to form two distinct brain types: female and male. An even broader question is whether sex differences in psychological characteristics add up to create two types of humans.

II

The Human Mosaic

5.

Brains in Flux

A close friend of mine—I'll call her Lisa—like myself, has three kids. "They are such girls!" she used to tell me when they were growing up. Ella, the oldest, a student of philosophy, is affectionate and dreamy; as a little girl, she could quietly keep herself busy for hours, telling stories to her veggies before eating them. The second daughter, Odelia, a terrific fixer of all things broken, is a born caregiver: while in kindergarten, she used to look after younger children, buttoning up their coats before they went home. Lisa's third daughter, Andrea, was so daring as a child, she'd started climbing sliding boards in the playground before she could walk; she is a wizard with cosmetics, loves to dress up in fanciful clothes, and is planning on becoming an actress.

In describing her daughters as "typical girls," Lisa picked a different "girl" trait for each kid. She is also well aware that each of her daughters has typical "boy" traits, again different ones in each case. So each of Lisa's kids has a unique mix of "feminine" and "masculine" traits, as do most people.

Given such variability, can we divide people into two types: "female" and "male"? We can certainly divide most people into female and male according to the form of their genitals, but does this division extend beyond the genitals, to their brains and personalities?

Each genital organ, internal or external, almost always comes in one of two distinct versions, female or male: clitoris or penis, labia majora (the outer lips of the vulva) or scrotum. And these organs typically add up *consistently* in each person. Having a penis usually goes with having a scrotum and other male sex organs, but not a vagina. A vagina usually comes together with having a womb and other female genital organs, but not a penis.

Now imagine that thirty minutes of stress changed your labia majora into a scrotum, or your scrotum into labia majora. Or that females were born with a penis instead of a clitoris if their mothers had been stressed during pregnancy, whereas males were born with a clitoris instead of a penis if *their* mothers had been stressed during pregnancy. In such a world, people wouldn't think of the labia majora as a female organ or of the penis as a male organ, because these organs would appear in both sexes. Moreover, talking about female and male genitals would make no sense, because many people would have various combinations of sex organs, not just the two most common. (I say "most common" because there are more than two types of genitals—about 1 percent of people have intersex genitals, so called because they don't fit into either the male or the female category—but most people have a set of genital organs clearly distinguished as male or female.)

Genitals don't change in this manner in humans. Our mother's stress during pregnancy or our own stress at any age certainly does not flip the form of our genital organs from male to female, or vice versa. However, stress does flip the "sex" of certain features in the brain.

That's precisely the amazing discovery I mentioned in chapter 1, the one that transformed my thinking about sex and the brain. I read about it in a 2001 issue of the *Journal of Neuroscience,* which reported on a study in rats conducted by Tracey Shors and colleagues at Rutgers University.[1] This study showed that relatively brief exposure to stress can switch certain brain features from a male to a female form, or from female to male. Moreover, the researchers found that the exposure to stress had different effects on different brain features.

The Rutgers scientists investigated pyramid-shaped neurons in the hippocampus, a brain region that, among other functions, plays a central role in memory and spatial perception. These neurons have delicate tree-like extensions, called dendrites, at the pointed tip (apex) of the neuron body and at its wide base. The researchers noted that, compared to male rats, in female rats the "top" (apical) dendrites had more dendritic spines—tiny protrusions that receive input from other neurons. But when the researchers exposed rats to thirty minutes of stress and examined their brains twenty-four hours later, they saw something surprising.

The "sex" of the top dendrites changed under the influence of stress. Those of males were now bushy, assuming the shape that had originally been seen in females; the top dendrites of females, on the other hand, now had sparse dendritic spines, looking just like the dendrites that the males had had before exposure to stress.

The "bottom" (basal) dendrites, those at the base of the neuron body, also responded to stress, but differently. As long as the rats enjoyed a peaceful, stress-free life in the lab, there was no difference between males and females in this brain feature. But exposure to thirty minutes of stress increased the density of the bottom dendritic spines in males, whereas in females, the density remained the same. This means that the exposure had *created* a sex difference that wasn't there earlier.

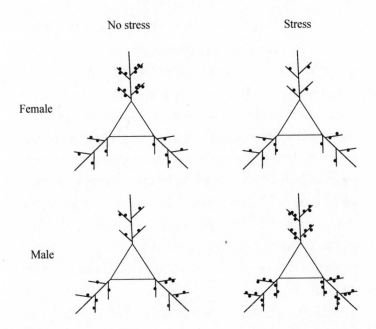

A schematic illustration of a rat's pyramid-shaped neuron: body (the large triangle), top (apical) and bottom (basal) dendrites (lines extending from the body), and the dendritic spines (the dots on the dendrites). The figure illustrates how the effects of sex (female or male) and of exposure to thirty minutes of stress combine to produce three types of neurons, those belonging to nonstressed females, to stressed females and nonstressed males, or to stressed males. *(Created on the basis of figure 4 in Shors et al., 2001)*

I was not at all surprised to read that neurons can change in response to stress. That's yet another example of neural plasticity I've discussed in chapter 4—the brain's ability to change in the wake of experience. What struck me was that the presence or absence of stress could have such a dramatic effect on the way an animal's sex affected the form of its neurons.

When I talk about the Rutgers study in public, I sometimes invite the audience to imagine that I'd just been extremely stressed on my way to the lecture because I was stuck in traffic, and that when I'd gone to the bathroom upon arrival, I'd discovered that my clitoris had turned into a penis. Some people laugh, others seem shocked at this prospect, but the imaginary analogy invariably gets across the point about the difference between brains and genitals.

The Rutgers study hadn't explored *what* it is about sex that affects the neurons—but something certainly does. It could, for instance, be the presence of an XX set of sex chromosomes in the neurons of females versus an XY in the neurons of males, the levels of such sex-related hormones as testosterone, sex differences in other hormones or other variables, or a combination of any of these.[2] But the researchers did show that *how* sex affects a brain feature may depend on other factors. The rats' being male or female produced different outcomes in their neurons depending on whether they had been stressed or not.

No less surprising was the observation that stress didn't "flip" the entire neuron from a "male" form to a

"female" form, or the other way around. Rather, the interaction of sex and stress led to the emergence of three forms
of neurons: those with sparse "top" and "bottom" dendritic spines (in stressed females and nonstressed males);
those with bushy "top" and "bottom" dendritic spines (in
stressed males); and those with bushy "top" and sparse
"bottom" dendritic spines (in nonstressed females).

In other words, such a simple manipulation as sub-
jecting an animal to thirty minutes of stress not only
changed the "sex" of a single neural feature but also
created three patterns of spine density that could not be
neatly sorted into "male" and "female."

Shors and colleagues followed up with another study,
which showed that stress can also reverse the way sex affects brain function—more specifically, learning.[3] They
found that under normal lab conditions, female rats
acquired a Pavlovian eyeblink response to a tone faster
than did male rats. However, just twenty minutes of
stress reversed this sex difference: male rats now acquired
this response as fast as did nonstressed females, whereas
females now learned at the same slower pace as that of
nonstressed males.

I was so captivated by these revelations that I wanted
to find out whether they were the exception or the
rule. So I started looking for more studies with similar
findings.

They weren't hard to find. The earliest one, to the
best of my knowledge, performed by Janice Juraska at

the University of Illinois, had been published as far back as 1985.[4]

As I was reading one study after another in which the effects of an animal's sex on the brain were shown to differ under various conditions, I began to realize that in thinking about sex and the brain, the body system people have in the back of their minds is the genitals. In connection with sex, that's the system we know most intimately, in every meaning of the word. That is probably why it has become the mold that shapes our entire perception of the effects of one's sex on the rest of the body. But I also began to realize that the logic that is familiar to us from the genitals does not apply to the brain.[5]

Consider one major difference between the brain and the genitals that I've already mentioned. Each genital organ, internal or external, almost always comes in one of two distinct versions, female or male. But we've just seen that even a single neuron can come in at least three forms, and this is also true of entire brain regions.

Margaret McCarthy and colleagues at the University of Maryland showed that in rats, chronic stress alters the "sex" of a particular feature in the hippocampus: the density of cannabinoid receptors (they are activated by a class of molecules produced by the mammalian body and by the cannabis plant).[6] If rats go about their stress-free routines, males usually have lots of these receptors throughout the hippocampus, and females have relatively few. This looks like such a simple division, it's tempting to talk about a "male" and a "female" pattern of cannabinoid receptors in the hippocampus. But let stress into the picture, and the result is anything but simple.

The Maryland scientists showed that three weeks of mild stress reversed this sex difference in the upper part of the hippocampus. Male rats acquired a "female," low-density pattern of cannabinoid receptors, whereas females acquired a "male," high-density pattern. The same stress exposure produced different results in the lower part of the hippocampus: it reduced the density of cannabinoid receptors in males but had no such effect in females, so, in effect, the stress exposure *erased* the preexisting sex difference—both male and female rats now had the low-density, "female" pattern of cannabinoid receptors in this brain region. As a result, in terms of cannabinoid receptor density, the hippocampus can come in at least three forms, each combining different "male" and "female" receptor patterns in its upper and lower parts.

You may be asking yourself by now: if it's so easy to reverse, create, or erase what is typical of males and what is typical of females, what's the meaning of talking about the sex of one brain feature or another to begin with? You would be absolutely right. I, too, have grown convinced that talking about the sex of brain features is meaningless.

Take one of the largest and most frequently cited differences between the brains of women and men: the size of a deep-seated cluster of neurons called the intermediate nucleus of the hypothalamus.[7] This nucleus is, on average, twice as large in males as in females, but this is true only until middle age, during which the nucleus starts shrinking in men, eventually reaching a size typical of women. In another cluster of neurons (which goes by the impossible name "the bed nucleus of the stria terminalis"), also known to be larger in males, this difference only appears in adulthood.[8] So what's the "true" male form of these nuclei, the one typical of children, young adults, or the elderly?

Rather than referring to "male" and "female" versions of brain features, wouldn't it make much more sense to use informative terms such as dense versus sparse, long versus short, big versus small? I'm not saying that sex doesn't matter. It certainly does, as the examples above demonstrate. But this doesn't mean that any particular brain feature or the brain as a whole can be assigned a sex category.

I hope that by now it is obvious that when applied to the brain, the logic familiar to us from the genitals falls apart on at least three fronts. First, there's the fact that human genitals typically remain fixed in their forms throughout a person's lifetime, whereas human brains do not. Second, genital organs almost always come in one of two distinct versions, female or male, whereas brain features can come in more than two forms. Third, genitals typically come in a set: most people have either female genital organs alone or male genital organs alone; most brains, in contrast, are mosaics of "female" and "male" features.

If we do persist in applying to brains the same terminology we apply to the genitals, we are bound to conclude that most brains are neither male nor female— they are intersex. Genitals are termed intersex when one or more genital organ is intermediate in form between typically female and typically male, or when one has both typically male and typically female genital organs. As I mentioned, people with intersex genitals are rare. But should one assign a sex category to brains, the majority would be classified as intersex.

6.

Not by Sex Alone

In contrast to the genitals, where sex plays a dominant role in determining the shape of each organ, in the brain, sex is only one of many determining factors. Stress is another, as we've seen in the previous chapter; many studies investigate stress because it affects the brain in important ways—for example, it's a known contributor to brain pathology. But numerous other factors interact with the effects of sex in shaping brain features: living conditions, upbringing, exposure to drugs, and other external influences.[1]

These influences—similarly to stress—may have different effects on sex differences in different brain features. Janice Juraska, for example, checked whether the same sex differences would show up in the brains of rats brought up in contrasting environments—in one group, each rat was kept in isolation in a standard cage; in the other group, rats lived together with other rats of the same sex in a cage outfitted with objects that were changed every day.[2] The scientists focused on brain areas involved in cognitive functions, including several regions of the cortex, the hippocampus, and the corpus callosum, the bundle of fibers that connects the two brain hemispheres. They found that whether higher measurements were obtained in females or males—and whether sex differences existed at all—was influenced by the environment. For example, in rats reared in isolation, the length of dendrites (those tree-like extensions of neurons) in the visual cortex was the same in both sexes in layer IV (the visual cortex, like the rest of the cerebral cortex, is organized in layers), but greater in females than in males in layer III; in contrast, in rats reared in the complex environment, the sex difference in layer III was reversed (the dendrites were longer in males than in females), whereas in layer IV, a sex difference had emerged (males now had longer dendrites than females).

Similarly, a drug might have different effects on sex differences in different brain functions, which manifest themselves in behavior. In one study conducted in my lab, my team explored, in rat pups, the effects of Prozac, a drug commonly used by pregnant or breast-feeding women.[3] Note that Prozac is known to affect behavior when given to adult rats, but in our experiment we gave it to pups in their first week of life and compared their behavior to a control group three months later, when they'd reached adulthood.

We found that early exposure to the drug altered the behavior of females and males in different ways for different behaviors. It increased depression-like behavior in females but decreased it in males, which, in effect, *reversed* the sex difference found in controls (control males had higher levels of this behavior than females). In other words, what was typical of females under one set of conditions became typical of males under a different set, and vice versa. On the other hand, anxiety-like behavior increased in females to the level of control males but stayed the same in males, thus *erasing* the sex difference (here again, control males had higher levels of this behavior than control females).

The complex interactions between sex, stress, drugs, age, and many other factors can determine numerous aspects of brain structure and function. In the previous chapter, we considered the effects of a single event at a specific point in time on only two brain features— "top" and "bottom" dendrites or the density of cannabinoid receptors in the upper and lower parts of the hippocampus—and ended up with a brain cell or region that has three forms. Now multiply these examples by the many types of brain cells and brain regions, and the numerous types of neurotransmitters and receptors. Then multiply them again by the enormous variety of events each of us has lived through from the moment of conception, and you'll understand why it is hard to imagine that brains can be consistently "male" or "female."

That's precisely why I argue that although sex does affect the brain, there are no "true" male and female brains out there to discover. The true nature of the brain is that its form is highly variable—a variability created by the interaction of multiple factors, including sex, in the fetus and throughout a person's life.

Now, what about the big "T"? Isn't there plenty of evidence that already in the womb, testosterone affects the form of the male baby's brain, turning it from female to male?

The story about a testosterone surge acting on the brain of the male fetus is not untrue; it's just not the whole story. If testosterone were the only factor responsible for the "sex" of brain features, and if it acted via a single mechanism on all these brain features, then we would expect the entire brain to be consistently male under high testosterone levels, or to remain consistently female under low testosterone levels. People with intermediate levels of this hormone would have brain features that are consistently intermediate between female and male, and brains would be aligned on a female-male continuum according to the degree of their exposure to testosterone.

However, in reality, none of these assumptions fit what scientists currently know.

Testosterone does indeed play an important role in determining the form of brain features, but other hormones, including estrogen, also affect the brain. For example, a study conducted in the laboratory of Janice Juraska has found that, compared to male rats, female rats have a lower number of a certain type of brain cell, known as glia, in the upper layers of the prefrontal cortex—but this is true only if their ovaries are intact.[4] Female rats whose ovaries had been removed had the same number of glial cells in this region as did male rats. Castration of the males, on the other hand, did not affect the number of these cells. These observations reveal that the sex difference in this brain feature in normal rats results from ovarian hormones acting on the brains of females, rather than from testosterone, or some other testis-derived substance, acting on the brains of males.

So there are numerous sex-related hormones that affect brain structure, not just testosterone. Moreover, the meticulous work of Margaret McCarthy and other researchers has revealed that these hormones act in the brain via multiple mechanisms, may have different effects in different tissues, and, as we have seen, may have completely opposite effects under different conditions.[5]

See what happens if we add only one ingredient— stress—to the story of testosterone and the male fetus.

During the many weeks of pregnancy, the fetus's mother sometimes gets stressed. And whenever this happens, some features of her fetus's brain change their "sex." So when her baby boy is born, his brain is already a mosaic of "male" and "female" brain features. This mosaic is uniquely his, molded by the complex interactions between his genes and hormones and the environment in which he had been developing. The same happens in the female fetus. Her brain is also molded by the complex interactions between genes, hormones, and the environment, so that the baby girl is born with a brain composed of a unique mosaic of "female" and "male" brain features.

People who learn about my work sometimes mistakenly conclude that I'm claiming there are no sex differences in the brain. I hope by now to have made it clear that's not the case at all. Sex does affect the brain, and there are average differences between females and males in many brain features. But because of the interactions between sex and so many other factors, the effects of sex—that is, of being female or male—mix up in a unique way in the brain of each individual.

If sex differences did add up consistently in our brains, then we could liken women's and men's brains to the make of cars. Car makes differ in numerous ways— in the engine, body style, seating, and so on. Some of these differences might be small, but taken together, they add up to create distinct types of cars. That's the analogy used, for example, by neuroscientist Larry Cahill from the University of California, Irvine. He wrote in an article in *Cerebrum:* "Claiming that there are no reliable sex differences on the basis of analyzing isolated functions is rather like concluding, upon careful examination of the glass, tires, pistons, brakes, and so forth, that there are few meaningful differences between a Volvo and a Corvette."[6]

But what if various car parts changed under different conditions? Each car would then end up with a unique combination of parts—and it would make no sense classifying cars into makes. My colleagues and I pondered such a scenario in our response to Cahill in *Cerebrum.* "Would we classify glass, tires, pistons, brakes, etc. as being of Volvo or Corvette origin if engines of Volvos changed form to become powerfully Corvette-like under some conditions, and trunks of Corvettes changed to become more spacious, depending on the specific social context in which the car found itself?" we wrote. "Or if, in some social contexts and countries, the pistons of Volvos differed quite significantly from those of Corvettes, but in other circumstances or countries they were the same?"[7] Obviously, this never happens with car parts—or genitals, for that matter—but as we've seen, it does happen with sex differences in the brain.

I had initially arrived at these conclusions about sex and the brain by reading dozens of studies performed in rats. But because each of those studies had looked only at a few brain features, I couldn't help but wonder if the male-female mosaic applied to the entire brain. Most of all, I was eager to know whether it applied to the entire brain of human beings.

7.

Mosaic of the Human Brain

Once, when I was a child, my family was stranded in the desert during a weekend outing because a pipe in the car engine had leaked. My dad saved the day by having us all chew gum, which he then pasted onto the leak. I was so impressed by his mechanical skills that after getting my driver's license, I volunteered for several months to work one morning each week in a nearby garage, changing engine oil and calibrating the timing of the pistons, so as to learn to fix my own car if need be. I enjoyed it so much that when submitting my application to university later that year, I even considered studying mechanical engineering. While waiting for school to start, I traveled to the

United States, got accepted to a large international modeling agency in New York City, and for a while tried my luck as a fashion model. Eventually, I didn't follow either the car mechanic or the modeling route, opting instead for medical sciences and then neuroscience. But looking back at these early experiences, I see that my own mix of "feminine" and "masculine" interests had revealed itself long before I became intrigued by the way "female" and "male" features coexist in our brains.

When I decided to examine this question in my research, it occurred to me that magnetic resonance imaging (MRI), one of the most popular methods of studying the human brain, would be perfectly suited for this task, because MRI scans provide information about the structure of the entire brain.

My plan was to identify the brain features exhibiting the largest differences between women and men, to define the "female" and the "male" forms for each feature, and then to test how many brains were consistently "female" or consistently "male" and how many had both "female" and "male" features. I asked Yaniv Assaf, an expert in structural imaging from Tel Aviv University, to join me in this project. Together with a team of students, we measured the volumes of 116 gray matter regions in a preexisting dataset of brain scans of 281 Israeli women and men.[1]

There were many differences, on average, between the sexes, but the overlap between the volumes of each region in women and men was so great that it was impossible to define a "female" and a "male" volume even for the regions with the largest differences. So instead of dividing volumes into "female" and "male," we divided the range of scores for each region into three categories. We called "female-end" the scores that were more common in women than in men; "male-end," the scores that were more common in men than in women; and "intermediate," the scores in between these two (they appeared with similar frequency in men and women). For example, if a region was larger, on average, in women, then the range of scores in the third of the women with the highest scores was defined as "female-end," the range of scores in the third of the men with the

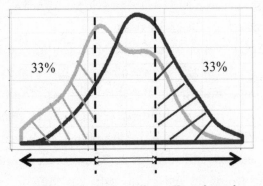

Male-end   Intermediate   Female-end

Distribution of the gray matter volumes of the left hippocampus in men (light gray) and women (dark gray). The "male-end" zone covers the scores for the 33 percent of the men with the smallest volumes. The "female-end" zone covers the scores for the 33 percent of the women with the largest volumes. *(Reproduced with permission from figure 1 in Joel et al., 2015)*

lowest scores was defined as "male-end," and all the scores in between were "intermediate."

We then selected ten regions with the greatest differences between the sexes. We didn't concern ourselves with the origin of these sex differences—that is, whether they came from nature or nurture. We simply selected the largest ten and went back to our 281 brains to check, for each of them, how many of the ten brain regions fell in the "female-end," "male-end," or "intermediate" range.

Having seen how the "female" and "male" features were mixed up in the brains of rats, I had expected that at least some of the human brains would be mosaics of "female-end" and "male-end" regions. But when we were done analyzing the data in humans, the results exceeded all my expectations. A mere 7 of the 281 brains—about 2 percent—had only "female-end" or only "male-end" scores; another 4 percent or so had only "intermediate" scores. The rest were mixtures of "intermediate," "female-end," and "male-end" features. Each brain had its own unique mixture, but most amazingly, about a third of the brains had features on both extremes, the "female-end" and the "male-end"— that is, they had extreme scores "appropriate" for their own sex on some of the features, but at least one score "highly inappropriate" for their sex.

I was, of course, thrilled that my hunch had proved accurate, but what if the results were true only for one particular collection of brains? We wanted to see if they applied to larger samples of brains and whether they were valid for other types of MRI and other methods of image analysis. We created a collaboration with neuroscientists from the Max Planck Institute for Human Cognitive and Brain Sciences in Leipzig and from the University of Zurich, and performed a similar analysis on MRI scans of brains from three additional datasets. All together we analyzed more than 1,400 human brains. We looked not only at the volume of gray matter (where the neurons reside) but also at the volume of white matter regions (which harbor the fibers connecting neurons) and the overall thickness of different cortical regions.

The regions showing the largest differences were not the same in the Israeli and the international collections of brains—just as could be expected, considering that life's conditions, which affect sex differences, are not the same in different parts of the world. Yet, regardless of which regions showed the largest sex differences, in all these analyses the same pattern emerged. The number of brains with only "female-end" or only "male-end" features was small; it ranged from 1 to 8 percent in the different datasets. On the other hand, the number of mosaics with both sex-"appropriate" and sex-"inappropriate" features was between 23 and 53 percent, depending on the dataset. Not only that, but among this latter group, some women had more "male-end" than "female-end" features, whereas some men had more "female-end" than "male-end" features.

We then examined, also with MRI, the strength of the brain's wiring—connections between different gray matter regions. One often hears that women's and men's brains are "wired" differently; it's not just the general public but scientists, too, who occasionally make this claim.

In a study published in 2014 in the *Proceedings of the National Academy of Sciences, USA,* for example, researchers at the University of Pennsylvania scanned the brains of 949 young men and women and found that the men, on average, had stronger connections *within* each hemisphere, whereas the women, on average, had stronger connections *between* the two hemispheres.[2] Because these researchers assumed that the average differences added up *consistently* within each brain, they concluded that their findings reflected two types of connectivity, male and female. One of the research team members even drew far-reaching conclusions from these findings, telling *The Guardian* that "the greatest surprise was how much the findings supported old stereotypes, with men's brains apparently wired more for perception and coordinated actions, and women's for social skills and memory, making them better equipped for multi-tasking."[3]

Our study offered an opportunity to put the assumption about male and female types of brain wiring to a test. We examined more than 4,000 connections in the brain. Like the University of Pennsylvania researchers, we found average differences between men and women in the strength of some of these connections. But then we went further. We looked at individual brains to test whether within each brain, these differences added up consistently. As in our previous analyses, we chose to focus on connections that showed the biggest differences between the sexes—in this case, seven of them.

Practically all the brains proved to be patchworks in terms of the strength of their connections. In a tiny fraction of the cases, 0.7 percent, all seven connections were of an "intermediate" strength, but we did not find a single person in whom the seven were all "female-end" or all "male-end." On the other hand, the number of brains containing connections on *both* extremes was very high: 48 percent of the entire lot. Because of this mix, it was impossible to talk about two types of connectivity, one typical of females, the other typical of males. Instead, typical of both women and men was a mosaic of connections: some with a strength more common in women, others with a strength more common in men, and yet others with a strength common in both women and men.

To conclude our study, we looked at the entire brain—really looked—in color. We entered the scores for the volumes of the 116 regions of gray matter in each brain into two tables—one for women, the other for men—but instead of numbers, we presented each region in each brain using a continuous high-to-low (green-white-yellow) color scale. This scale represents the volume of a brain region in a given brain relative to the volume of this region in all other brains. If a region was relatively large, we assigned it a shade of green (the larger the region, the darker the green). If a region was relatively small, we assigned it a shade of yellow (the smaller the region, the darker the yellow). Regions of an intermediate size were painted white. Each row represented a single brain, and each column stood for a different brain region.

When we saw the resultant tables—featured on the back cover of this book—we realized that the mosaics had materialized, quite literally, in front of our eyes. Overall, there was more green in the women's table and more yellow in the men's—that's because women have, on average, more gray matter than men relative to total brain size. Yet the vast majority of brains were not all green, all yellow, or all white. Rather, they were mosaics of regions: some green (relatively large), some yellow (relatively small), and some white (intermediate in size).

Here's what we reported on December 15, 2015, in the *Proceedings of the National Academy of Sciences, USA:* "Most brains are comprised of unique 'mosaics' of features, some more common in females compared with males, some more common in males compared with females, and some common in both females and males."[4]

People often ask me if our findings contradict the numerous studies that had found sex differences in brain structure or connectivity. My answer is that they do not.

Our study, like many others, found average sex differences in brain structure and connectivity. But what no study had done before ours was to check whether these differences showed up consistently within individual brains. We used the largest sex differences and asked whether they added up. What's new in our study is that we went beyond group-level sex differences—and found that they don't add up in a consistent way in individual brains to create two types of brain.[5]

It may sound contradictory. How can there be differences between women and men in brain structure, yet no female and male brains? Nothing probably clarifies this seeming paradox better than the table of results I've just described. I'd like to invite you to have another look at that table. You don't need to know much about statistics to notice that there are differences between men and women at the group level. But you can also see that brains do not come in two forms, green and yellow, but rather in multiple forms, each a unique mosaic of different shades of green and yellow, with a bit of white sprinkled over.

We can conclude that the human brain is neither

female nor male. Rather, it is a unique mosaic of features, some more common in females, others more common in males. This mosaic continues to change throughout our lives, like the ever-changing pattern of colored pieces in a kaleidoscope.

8.

Now You See It,
Now You Don't

The mosaic notion can help explain the Cheshire cat–like property of sex differences in the brain: their tendency to vanish now and then—that is, to show up in one study but not in another. I'd like to propose the following analogy. A cook has prepared a soup with lots of different-flavored noodles—chicken, beef, spinach, mushroom, and so on—and poured this soup into two large pots. The soup in this analogy is the human population, and the different-flavored noodles are the widely varied brains. Now soup experts who happen to walk into the empty kitchen wrongly assume that the pots contain two different soups. They are ladling soup samples from the two pots, trying to determine how the two soups differ by comparing the flavors of the noodles in each bowl. Unaware that in fact they are sampling the same soup, they come up each time with samples that differ in different ways. No wonder their results are inconsistent.

Something similar might be happening in studies of sex differences in the brain. Whenever a study reveals a difference between women and men, particularly in brain function, almost invariably, at least one follow-up study fails to find a sex difference in brain activity in the same regions. We saw in chapter 3 that this happened with sex differences in activity patterns during language processing, and those conflicting results weren't an exception by any means. In fact, the most consistent finding in studies of sex differences in brain function is that their results tend to be inconsistent.

This was the case with findings from several studies that compared brain activity in men and women as they performed mental rotation. In this task, people are usually shown a drawing of a three-dimensional object and are asked to compare it with drawings of similar objects, to determine which of these represent the original object after it's been rotated. Men outscore women, on average, in this task.

In one study, published by German scientists in 2002, men performing mental rotation had higher activation in some regions of the frontal lobes, whereas women performing the same task had higher activation in some regions of the parietal and temporal lobes.[1] But a year later, another study contradicted some of these results. Austrian scientists found that the same parietal region that had been shown in the German study to be more active in women was more active in men in both hemispheres.[2] A third study, performed in 2006 in Norway, found a greater activation in men in the same parietal region as in the Austrian study in the right hemisphere, but failed to replicate this finding in men's left hemisphere.[3] A fourth study, conducted the same year in the United Kingdom, found no differences at all between brain activity of women and men as they performed mental rotation.[4] And a fifth study, conducted in 2010 in Spain, found that many regions in the frontal, parietal, and occipital lobes were activated by the task, but there was a difference between women and men in two occipital regions only.[5] Talk about a Cheshire cat: this sounds like an entire bunch of them.

Mental rotation—like other cognitive functions—involves the participation of an entire network of neural regions spanning different parts of the brain. If a study of a specific cognitive function reports on sex differences in some of these regions, and another study of the same cognitive function reports on differences in other regions of the network—as in the case of the mental rotation studies—one might conclude that the two studies support each other's conclusions about the existence of sex differences in brain activity during this task. But as Rebecca Jordan-Young, of Barnard College, Columbia University, explains in her book *Brain Storm,* such reports, in fact, contradict one another, each failing to replicate the sex differences found in the other study.[6]

Many scientists have been puzzled by this lack of consistency. It runs so counter to their expectations that they've tried to blame it on the inclusion of too few participants in some of the studies, on inadequately defined tasks, or on faulty statistical analyses. These are all valid possibilities. But the one thing conspicuously lacking in all the explanations has been the suggestion—which should form a basic tenet of scientific exploration—that if research findings fail to support a theory, then perhaps it's the theory, rather than the research methods, that needs to be revised. Apparently, the notion of fundamental differences in brain and behavior between women and men is so embedded in our culture that giving up on this idea is unthinkable for many people.

But if there are no such fundamental differences, why is it that some of the studies do find differences in brain activity patterns between women and men? I'd like to argue that in many, if not most, cases, this happens simply by chance and is due to the huge variability of human brains and the typically small size of the samples.

A mathematical approach called statistical inference is intended to prevent this from happening. It provides tools to help researchers decide whether differences between two samples—let's say, between two collections of brain scans—are likely to be the result of mere chance, or of each sample representing a different type—in this case, a different type of brain. Statistical inference works better with larger samples. This means that if brain activity patterns indeed fall into two types, female and male, then studies with larger numbers of brains should discover more sex differences than those with smaller samples. But an analysis of 179 studies comparing brain activity patterns in women and men, published in 2018 in *Scientific Reports,* revealed no such relationship.[7]

The mosaic view of the brain explains why studies comparing the brains of women and men produce such inconsistent results: why many fail to find any differences, and why those that do find differences report conflicting findings. As in the soup analogy, the studies are sampling brains from a *single* hugely variable human population— each time coming up with different differences between the brains in their samples.

9.

In Anticipation
of a Blind Date

In *RBG*, a documentary about the life of Ruth Bader Ginsburg, the US Supreme Court justice recalls that in her first semester at Cornell University, she "never did a repeat date." Things changed when she met her future husband, Martin D. Ginsburg. "He was the first boy I ever knew who cared that I had a brain," she quips.

That was back in the early 1950s. Studies show that today, men increasingly desire women who are educated and intelligent and not necessarily good at cooking.[1] But can a person seeking a partner—of any sex—know what kind of intelligence and cooking abilities he or she is likely to possess?

Picture yourself going on a blind date when all you know about the prospective mate is that it's a female or a male. This provides you with much information about your date's genitals, but what else can you tell in advance? Does your date's sex tell you what kind of brain, and what kind of personality, preferences, or attitudes she or he has?

You won't find the average differences between the brains of females and males of great help, even though these differences will most often allow scientists to correctly guess whether a given brain scan comes from a man or a woman. (By the way, this latter ability does not contradict my claim that there's no such thing as female and male brains, as I'll explain in this chapter.) But guessing the sex of a brain's owner is not particularly interesting—that's something we typically know by looking at that person, and in case of a doubt, it is much easier to obtain that information by simply asking the person, or by looking at her or his genitals, than by assessing her or his brain scans. The really interesting question is just the opposite: Knowing that we are about to look at the brain of a male or a female, can we tell in advance what kinds of features we are likely to encounter?

Our study has shown that we cannot. Knowing someone is male, for instance, tells us very little about his brain. The only thing we can say is that brains of males are likely to have more features that are common in males than those that are common in females. But we have no way of knowing how many of each, and which ones.

Remember my friend Lisa who said that her three daughters are "typical girls" because they all have certain "girl" traits? She couldn't tell in advance what kind of traits each of them would have. Even after having one girl, she was in for a surprise each time she gave birth to yet another. Similarly, when you go on a blind date and all you know is that your date is male or female, this hardly tells you which masculine or feminine features to expect.

Consider the following hypothetical example of a brain that has only three regions. Let's say we label the three brain regions with the letters *A, B,* and *C* if they come in the versions that are common in females, and by *a, b,* and *c* if the same three regions appear in versions that are common in males. So if you encounter someone with an *ABC* or an *ABc* brain, this person is probably a female, and if you encounter an *abc* or an *Abc* brain, it most likely belongs to a male. But if you encounter a female—on that same blind date, or under different circumstances—can you tell in advance what kind of brain features she has? The answer is that you *can* predict she is likely to have more features that are common in females than those common in males, but you cannot predict the exact composition of her brain mosaic: it might be *ABC, aBC, AbC,* or *ABc.* And, of course, she may be one of the relatively few females who have more features that are common in males, so she may have an *Abc, aBc, abC,* or even an *abc* brain.

The three-region example can also help explain why, when it comes to deciding whether two brains are similar, the exact composition of their mosaics is more important than the proportion of "feminine" or "masculine" features they contain. For example, an *aBC* brain, which is supposedly more feminine, is more similar to a supposedly more masculine *aBc* brain than to another "feminine" brain, *AbC* (because *aBC* and *aBc* have two feature versions in common—*a* and *B*—whereas *aBC* and *AbC* have only one—*C*).

When we move from a hypothetical brain with three regions, each of which has only two forms, to the real human brain, the number of potential mosaics skyrockets. Not only can you not predict a person's brain mosaic on the basis of their genitals, but this information is also useless for predicting whether their brain is similar to yours or to anyone else's. That's just what a collaborative study of mine has shown, as I'll describe in the next chapter.

10.

Brain "Types," Common and Rare

Whenever I talk about the brain mosaic, one of the objections people raise is the different rates in males and females of such brain-related disorders as autism and depression. Some people interpret these differences to suggest that they reflect the nature of typically male or typically female brains.

For example, Simon Baron-Cohen, of the University of Cambridge, claims that autism is an extreme form of the male brain.[1] He assumes that there is such a thing as a male brain—a *typical* male brain, not an autistic one—and that the ways in which it differs from a typical female brain make it a little more autistic, so that in extreme cases, the brain's owner has autism.

I disagree. Autism is a rare condition. The fact that there are more males than females among people with autism does not necessarily mean that typical males are a little autistic. Similarly, the fact that there are more females than males among people suffering from depression does not mean that typical females are a little depressed.

On the other hand, it is surely plausible that specific brain mosaics—which are yet to be discovered—related to autism or depression are more common in males or in females. Brains consisting of "male-end" features only, although rare, are nonetheless more common in males than in females. The same is true for brains consisting of "female-end" features only: few people have such brains, but those who do are for the most part females. Therefore, the prevalence of brain-related disorders in one sex or the other may very well reflect sex differences in the prevalence of rare types of brain mosaics, rather than differences between supposedly typical male and female brains.

I have recently tested this hypothesis in collaboration with researchers from my university's School of Mathematical Sciences. In this study, published in 2018 in the journal *Frontiers in Human Neuroscience,* we employed several mathematical approaches to analyze the volumes of gray and white matter regions in the brains of 2,176 women and men.[2]

First, we used an anomaly detection algorithm, originally developed to detect abnormal computer activity—

it "learns" the normal activity of a computer so that it can later sound an alarm when it spots something abnormal: a potential computer virus or worm. Instead of training the algorithm on computer activity, we trained it with half of the women's brain scans in our sample, so that it learned what a "normal" female brain is. We then let the algorithm examine the other half of the women's brain scans and a similar number of men's brain scans, and determine for each brain whether it was "normal." The idea was that if the typical female brain is different from the typical male brain, then the algorithm would mark the brains of many more men than women as "abnormal."

It didn't. Rather, the algorithm marked as abnormal almost identical numbers of men's and women's brains it hadn't seen earlier. That was also true when we reversed the study, first training the algorithm on half of the men's brain scans in our sample, then asking it to examine men's and women's brains it was encountering for the first time. In effect, what happened in the study was that the algorithm defined common brain types for one sex—the "normal" brains—and subsequently revealed that they are also common in the other sex.

But what about my hypothesis that there are sex differences in the prevalence of rare brain types? To address this question, we employed an additional approach, called unsupervised clustering, in which an algorithm groups objects together based on similarity. Two such algorithms were given information on the structure of brains from both women and men, but not on the sex of each brain's owner. The algorithms created several clusters: some large (representing brain types common in humans), others small (representing rare brain types).

We found that in the large clusters, the numbers of brains belonging to women and to men were very close. This finding supported the results obtained with anomaly detection—that brain types common in women are also common in men, and vice versa.

Likewise, in many of the small clusters—that is, rare brain types—the number of women and men was similar. But in some of these small clusters, the number of brains from one sex was larger than the number of brains from the other—in certain clusters, up to six times larger. Since we limited ourselves to mathematical analysis of brain structure without going into the biological significance of one brain pattern or another, we couldn't tell how these patterns translate into brain function. But this finding did supply indirect support for my hypothesis that sex differences in the rates of certain psychopathologies could be related to the different prevalence of certain rare brain mosaics in males and females.

Finally, we calculated the chances of any two brains—from persons of the same or different sexes—being found in the same cluster. We discovered that a woman and a man had about the same chance of finding themselves in the same cluster as did two women or two men. This finding provided support from real brains for the theoretical claim I've made above, that information about a person's sex is useless for predicting whether his or her brain is similar to anyone else's.

Some people are puzzled by this lack of predictability. That's probably why Sarah Richardson, of Harvard University, who interviewed me for her GenderSci blog at Harvard's website, asked me to clarify: *How can it be that knowing the structure of a brain can often predict the sex of its owner, but knowing the sex of the owner is not very predictive of the structure of the brain?*

I suggested the following analogy, which was included in the GenderSci Q&A. Imagine that aliens come from outer space and want to describe human clothes to their home planet. They may come up with such categories as warm clothes versus light ones; large clothes (for adults) versus small ones (for kids); clothes you put on your legs, on your upper body, on your head, or on your feet; or they may create categories depending on color. If they are unaware of the social importance of the sex category, they may not create a category of men's clothes versus women's clothes. However, were they asked to classify clothes as belonging to a man or to a woman, they may be able to learn to do so, using such features as colors that are more common in the clothes of one sex over the other (for example, pink), designs, types of fabric or such trimmings as lace, and the direction of buttons in men's versus women's shirts.

Once they learn to make this distinction, they will be

able to use gender-related variability in human clothes to predict a person's sex on the basis of what he or she is wearing, just as we do every time we meet someone for the first time (in addition to relying on such attributes as hairstyle, accessories, and makeup). But this does not mean that the sex category reflects the most important aspects of the variability in human clothes. Knowing that someone is, say, male would enable the aliens to predict he most likely isn't wearing a lace outfit, but they wouldn't be able to predict whether he is wearing warm or light clothes. Similarly, as I told the GenderSci blog, "knowing that someone is a male is not informative about which brain structure he has, since sex is not a strong determinant of variability in human brain structure."[3]

To further clarify the prediction puzzle, here's yet another analogy, one that doesn't require imagining the arrival of aliens. Rather, it has to do with the use of language.

Computer scientists from Bar-Ilan University and Jerusalem College of Technology have designed an algorithm that can tell whether a given text was written by a woman or a man. They relied on findings that women's style tends to be more "involved"—that it works to establish a relationship between author and reader (for example, through the use of personal pronouns, as in the phrase "I think"), whereas men's style tends to be more "informational." The algorithm works by entering the number of occurrences of certain words and grammatical constructions into a formula that assigns a score to the text. If this score is greater than the total word count in the text, the author is most likely male; if it's smaller, the author is probably female. In one study, published in *Literary and Linguistic Computing,* the algorithm determined the author's sex with approximately 80 percent accuracy.[4]

But knowing the sex of the author provides no information about the frequency with which each of the stylistic features occurred in the text, and it surely gives no clue as to whether this text is short or long, a memoir or a romantic novel, compelling or soporific. In other words, although gender differences in writing styles can be used to predict whether a text was written by a woman or a man, the author's sex category provides no useful information about the text itself—just as knowing that a given brain belongs to a woman or a man tells us very little about that brain's characteristics.

11.

Men and Women
Under Stress

People sometimes ask me: Could it be that human brains can be divided into male and female not by their structure or function but by the way they respond to external factors such as stress? Indeed, in discussing the brain mosaic, I provided many examples of stress effects on the brains of rats, and how these effects may be different in females and males. Stress also affects the human brain, as well as human physiology and behavior, and one commonly hears that it may affect females and males differently. For example, the annual Stress in America surveys, conducted by the American Psychological Association, find that women and men differ in the physical symptoms they tend to develop as a result of stress, in expressing stress-related grievances, and in ways of coping with them.[1]

But are there really two types of response to stress, female and male? Once the results of my brain mosaic study came in, I wanted to look into this matter.

Again, I didn't concern myself with whether the responses of women and men were shaped by sex or gender—that is, whether they'd been altered by the gendered expectations on how to behave in stressful situations. Men may be expected to conceal their emotions or even put up a fight in order not to lose face, in accordance with a macho ideal, whereas in Western culture it's considered more appropriate for women to express insecurity or fear. Rather, my goal was to see whether in each individual brain, stress-induced changes were consistently typical of women or of men—or whether brains showed a patchwork of responses, some changes being typical of women and others typical of men.

It is usually hard to compare how the brains of women and men change in response to stress, because to address this question, one would need to scan numerous brains before and after a stressful event, yet scans conducted prior to such events are rarely available. But my colleagues and I managed to identify a group of people in whom "before" and "after" scans had been performed.

We learned that a team of neuroscientists from Tel Aviv University had completed a study in which they had scanned the brains of thirty-four female and male army paramedics at the beginning of their army service and about three years later. All the soldiers—females and males alike—had been exposed to at least one highly stressful event during that period: they witnessed grave casualties and treated patients with severe injuries in the course of a war.[2] True, it wasn't a perfect experimental setup, because the study design was not controlled in the same way it is with laboratory animals. And yet because one doesn't induce extreme degrees of stress in human beings for the sake of scientific inquiry, it seemed to be as good as it gets for a study in humans.

After establishing a collaboration with the colleagues who had performed the scans, we looked for brain regions in which the response to stress differed between men and women. We found that the volume of many regions had indeed changed at the end of that stressful period, but surprisingly, in most regions these changes were similar in the two sexes. I say "surprisingly," because in discussions of the stress response in women and men, it is usually the differences that are being highlighted.

Next, we focused on the seven regions that did change differently in women and men. For example, after the stressful army service, the volume of a part of the corpus callosum typically decreased in women but increased in men; the volume of the pericalcarine cortex (one of the visual regions in the brain) also typically increased in men but stayed the same in women; and the volume of the inferior parietal lobule (a brain region involved in visual perception, including the perception of emotions) typically decreased in men but stayed the same in women. We then examined the brains one by one to determine, for each individual, whether each of these regions changed in a way typical of males or females.

Only in one soldier out of the thirty-four had all of the brain regions responded to stress in a consistent manner; they all showed a female-typical response. In contrast, in the brains of twenty-five soldiers, some regions responded in a way typical of females, while other regions responded in a way typical of males. In the brains of the other eight soldiers, some regions responded in a way that wasn't typical of either females or males; the remaining regions either all responded in a way that was typical of females or all responded in a way that was typical of males. Our findings suggest that at least in what concerns brain structure, there is no such thing as a female or a male response to stress. Rather, each person reacts to stress with a unique mosaic of changes.[3]

In our study, we didn't try to establish how this mosaic comes about. Each soldier's unique response could have been the result of his or her genes, hormones, experiences, personality, and myriad other factors. And that's precisely the point I want to make. The effects of a person's sex interact with so many contributing factors that despite the average sex differences in the response to stress, it's impossible to tell in advance how each of the regions in one's brain will respond to stress just by knowing that this person is female or male. Nor is it possible to tell in advance, based on a person's sex alone, what to expect on a blind date—as we've already seen. As I discuss in further chapters, we also can't determine which medical treatments will work best for a given person, which toys one will prefer as a child, or whether someone is suited for a particular job.

Let us go back to the Stress in America surveys to see if they perhaps provide evidence for the existence of a female and a male response to stress. The press release for the 2010 Stress in America report, which devoted an entire section to gender, opens with the statement: "Men and women report different reactions to stress, both physically and mentally."[4] But a closer look reveals that, similarly to studies of sex differences in other domains, the 2010 report mainly highlights the differences between women and men and ignores the similarities.

The report states, for instance, that women are more likely than men to point to money as a source of stress: the numbers are 79 percent for women and 73 percent for men. While mathematically this is true, in reality both genders—more than 70 percent of women *and* men—are likely to point to money as a source of stress. The same applies to women being more commonly "stressed out" about the economy (the numbers are 68 percent for women and 61 percent for men) and men being "far more likely" to cite work as a source of stress (76 percent, compared with 65 percent of women). Even if these gaps are statistically significant, the emphasis on the differences inadvertently creates the impression that the sexes are much further apart in their responses to stress than they truly are.

The same applies to the statement that men are "more likely than women to say they do nothing to manage their stress." The number for men is indeed higher—9 percent, compared with women's 4 percent. But it would be equally true to point out that *both* genders—in fact, more than 90 percent of men and women—are *highly unlikely* to report that they "do nothing" in response to stress.

The spotlight on the differences ends up painting such a partial picture that it reminds me of an old Cold War joke about a race between two runners, one from the United States and the other from the former Soviet Union, in which the American won. In the Soviet Union, it was reported that the Soviet runner came in second, and that the American came in before the last runner. True, but surely gives the wrong impression.

And of course, there's no way of knowing from the Stress in America report whether various strategies that are more common in women consistently show up in each individual woman, adding up to a "female" response to stress, whereas strategies that are more common in men consistently show up in each individual man, adding up to a "male" stress response. My coauthor, a female, reports that she tends to spend time with friends and family to help her cope with stress, as did 54 percent of women in the survey, as opposed to only 39 percent of the men. But she does not manage stress by reading, which places her with the 66 percent of men who said they didn't employ this coping strategy, rather than with the 57 percent of women who said they did. She is therefore a case of a stress management mosaic.

Stress in America surveys provide information on an issue of tremendous importance to public health. According to their stated goals, they aim, among other things, "to examine the state of stress across the country and understand its impact."[5] It would be good to analyze the results of these surveys while taking the mosaic notion into account. If such an analysis checked for consistency in individual women and men, I'm willing to bet it would find that most individuals reveal themselves as mosaics of worries and stress-coping strategies, some of them more common in females, others more common in males. And if this is indeed the case, stress management is much more likely to be effective if it takes individual human mosaics into account, rather than trying to group humans into two distinct categories based on sex differences of one sort or another.

12.

—

The Human
Health Mosaic

Treating females and males as if they belong to distinct types can be bad for our health—not only when it comes to the effects of stress. This doesn't mean that people's sex needs to be ignored in health-related issues—just the contrary. It means that the binary approach to sex can obstruct biomedical research intended to make us healthier.

For many years, females were excluded from clinical trials as well as from entire areas of basic research.[1] In animal studies, that exclusion was caused in part by the fear that females, with their hormonal cycles, would muddle results. (Such fears have been shown to be invalid;[2] male animals have been found to be as variable as females in hormonal levels—as I'll discuss below—and on a wide variety of other physiological and behavioral measures.) Another reason was that once the practice of using animals of a certain sex is established, it's hard to change. Scientists who had long been using male laboratory animals—for instance, in neuroscience—feared that including female animals would alter the outcome. As a result, the original sex bias in research perpetuated itself for decades.

Inclusion of females in research is an essential part of trying to capture the great variability among human beings. It's a good thing, then, that medical authorities around the world have stepped in to correct the imbalance. Women now make up just over half of participants in clinical trials funded by the National Institutes of Health (NIH) in the United States,[3] and in 2016, the NIH required that all basic research studies, which precede testing in humans, must include both female and male animals. The problem is that the pendulum has swung too far. The justified attention to including both sexes in research is fostering the binary illusion—treating females and males as if they had distinct physiologies.

One of the mottos capturing the binary divide in health has been "every cell has a sex."[4] This statement refers to the fact that in most individuals, cells have either XX or XY chromosomes (there are other, less common combinations, such as XYY and XXX). But this motto is misleading; it even reminds me of eighteenth-century beliefs I mentioned in chapter 2—that sex extends into every body organ. We've already seen that, at least in the brain, cells come in many different forms, not just one typical of XX cells and another typical of XY cells. That's because our cells are influenced not only by the XX or XY chromosomes but by many other factors, including age, lifestyle, body composition, and hormones.

Hormones don't fit into two categories, either.[5] The entire range of sex-related hormones, so called because they are made by the sex glands—the ovaries and testes—is found in females and males alike. These hormones are produced by *both* ovaries and testes, and some of them are also manufactured by other tissues, such as the adrenal glands.[6] Not only that, in humans, the average levels of the three main sex-related hormones—estrogen, progesterone, and testosterone—overlap in females and males throughout most stages of life. There's no difference between girls and boys in these three hormones from the second year of life until adolescence. In adulthood, the average levels of estrogen and progesterone are similar in women and men, except for certain peaks characteristic of females, for example, before ovulation and during pregnancy. The average level of testosterone is higher in males, starting at adolescence, but for this hormone, too, there is an overlap between the sexes.[7]

Moreover, in each person, the levels of all the sex-related hormones fluctuate in response to internal and external factors. I mentioned in chapter 4 that in women and men testosterone levels are affected by engaging in competition. Parenting styles affect this hormone, too: for example, fathers who spend a great deal of time with their infants tend to have lower testosterone levels than those who don't.[8] In male mice, the daily ups and downs of testosterone are much larger than the changes in the levels of estrogen and progesterone during the estrous cycle of females.[9]

So even the sex-related hormones don't come in two distinct "pink" and "blue" sets, estrogen and progesterone in one, testosterone in the other. And a person's hormone profile—contrary to the form of one's genitals—is not a fixed characteristic, but is dynamic and reactive. None of this complexity is captured by the binary division into "female" and "male" hormones or "women's physiology" and "men's physiology."

The binary approach hinders the progress of medical research precisely because it misdirects the researchers' attention to sex and gender categories and away from discovering what *really* happens in the human body.[10] For example, if a difference is discovered between women and men in symptoms or in response to treatment, it's crucial to determine which genes or hormones are responsible for this difference, or whether it results from a multitude of additional factors on which women and men differ, on average, including height, weight, muscle mass, physical activity, jobs, and hobbies, to mention just a few.[11] Some of the health-related differences between women and men have to do with sex, others with gender. But regardless of their cause, these differences don't add up consistently in each individual woman or man, telling us all we need to know about "women's health" and "men's health."

Here's an example of how a binary sex label was attached too soon. In 2013, the Food and Drug Administration cut in half the prescription dosage of the popular sleep drug Ambien—for women, but not for men. The FDA found that 15 percent of women felt drowsy in the morning after taking zolpidem tartrate (the active ingredient in Ambien), compared to 3 percent of men.[12] Then researchers discovered that among the major reasons for the discrepancy between the sexes was the average difference between women and men in such parameters as proportion of fat to muscle.[13] Ambien now comes in "her" and "his" bottles—with a pink label for a low dose and a blue one for the original dose. But this binary approach means that some people—for example, overweight men who are not very muscular— might get the wrong dose. And it fails to take into consideration the 85 percent of women who did just fine with the original high dose.

In another example, physicians had believed for a while that a certain treatment for heart failure benefited women more than men. But a 2018 study published in the *European Journal of Heart Failure* has shown that, in fact, cardiac resynchronization therapy—in which electrodes are inserted into the heart to synchronize the function of its ventricles—works better in shorter patients.[14] Since women are, on average, shorter than men, this greater success was mislabeled as a "female" advantage, but as it turns out, short men might benefit from the therapy more than tall women.

The emphasis on sex or gender categories can sometimes even be fatal. When a disorder is labeled as a "man's" or a "woman's" disease, it can be misdiagnosed in the "wrong" sex. People tend to think of heart disease as a man's disease, yet it is the number one killer of women in the United States, far exceeding deaths from breast cancer.[15] Because for years heart disease was studied mainly in men, symptoms that are more commonly reported by women—such as back pain, nausea, and headache—are viewed as unusual, often delaying correct diagnosis and treatment in situations in which every minute counts. Men who develop symptoms that are more typical of women, instead of the classical "male" chest pain, are also at risk of delayed or erroneous diagnosis.

Researchers have started removing the gender label from a condition viewed as such a female problem that even its name left no room for men: postpartum depression. It turns out that fathers, too, sometimes get depressed before or after the birth of a baby. In a 2018 study, published in *JAMA Pediatrics,* researchers found that the numbers for the two sexes were surprisingly close.[16] Among the thousands of new parents who filled out their questionnaires, the percentage of those who met the criteria for depression stood at 5 percent for mothers and 4.4 percent for fathers. Screening both parents for depression "could be critical for ensuring the best possible outcomes for children and their families," the researchers wrote in their report.

Personalized medicine of the future will take into account a variety of variables, some of them related to sex and gender, coming up with an individual health recipe for each person. Getting rid of the gender binary—and appreciating instead the complexity of human physiology and of sex itself—is a critical step toward making this approach work.

13.

Mosaic of the Mind

Jane, a California native who lives in Israel, has shared with me her own take on the mosaic idea. As a teenager and then a young adult in the 1960s and '70s, she had grown up with the feminist movement. It had a strong presence on the University of California's Santa Cruz campus, where she earned her undergraduate degree. Still, she says that as a young woman she often felt "like a freak" because she didn't fit the feminine stereotype. Though she was gentle and shy, she wasn't a nurturing type, and she was good at math and at fixing things. After immigrating to Israel and settling on a communal farm—a kibbutz—Jane at first worked in the kibbutz sewing factory but then applied for a position as a maintenance worker, which included repairing the equipment in the milking facility. This was a much better fit for her. She filled the position successfully for seven years, then trained a man to take over.

After reading about my research on the brain mosaic, Jane told me it matched what she had known from her own life. "I recall feminist arguments over women and men being different or not. I am happy that the brain mosaic notion doesn't include the argument that there is no difference between the sexes—but rather that the differences mix in ways that are much more complicated than a simple division into male and female."

How prevalent is this "mixing" of masculine and feminine traits? My colleagues and I decided to look into this question using the same scientific method we had applied to the analysis of brain structure—as part of the 2015 study published in the *Proceedings of the National Academy of Sciences, USA* that I've already discussed.[1]

Note that I've moved from discussing brain structure to the realm of personality and behavior—that is, brain function. The relation between brain structure and function is complex. Usually you cannot draw a direct cause-and-effect line between this or that brain feature and a certain aspect of personality or behavior. But the description of human brain structure as a mosaic, rather than as a set of "male"-only or "female"-only features, fits in perfectly with the idea that each of us has feminine and masculine traits, and that human beings are rarely all feminine or all masculine.

To test this idea, we obtained datasets on psychological characteristics and behaviors of large numbers of people and asked: Do sex differences in these domains add up in a consistent way to create two types of humans, each with its own set of personality traits, attitudes, interests, and behaviors, or, as in brain structure, do they create mosaics of feminine and masculine characteristics? As in the study of brain structure, we didn't concern ourselves with whether the differences between males and females stemmed from nature or nurture. We simply looked for the largest sex differences in personality or behavior, and wanted to see whether they added up to create two types of human nature.

We soon discovered what University of Wisconsin–Madison psychologist Janet Hyde and others had already shown: that the scores obtained by women and men on most parameters psychologists can measure are much more alike than is commonly believed, and that even when average differences are found between the sexes, they are usually small.[2] For example, when we analyzed a dataset from a national survey of some 7,000 Americans that assessed personality traits and well-being (Midlife Development in the US, or MIDUS), we discovered that we could not use these data. On all but one of the fifty-six variables in the survey, the differences between men and women were nonexistent, or so small that defining "female-end" and "male-end" scores would have made no sense.

But we did find a few datasets in which the differences between the sexes were sufficiently large for our analysis. We picked three, which together covered personality traits, attitudes, interests, and behaviors of over 5,500 American youth. Following the same approach as in the brain structure study, we chose variables on which females and males showed the biggest differences, on average, and defined "female-end" (feminine), "male-end" (masculine), and "intermediate" values for each.

In the dataset of 570 adolescents from a study called Maryland Adolescent Development in Context Study (MADICS), we picked seven variables on which young women and men differed the most, including self-esteem, worries about weight, and the participants' communication with their mothers. There wasn't even a single person who had only feminine or only masculine scores, whereas 59 percent of the youth had both feminine and masculine characteristics. The table on the back cover speaks for itself: it presents each person in the study as a unique mosaic of different shades of pink ("female-end"), white ("intermediate"), and blue ("male-end").

As with the table of brains from chapter 7 that also appears on the back cover, you can easily see the differences at the group level, with more pink in the women's table and more blue in the men's. But when you look at each individual—that is, a single row—you see that most of them have both pink and blue features, while none have only pink or only blue.

Nor was there a single person with only feminine or only masculine characteristics in a much larger dataset, of 4,860 adolescents, from the National Longitudinal Study of Adolescent Health. The variables with the largest gender differences were depression, perceived weight, delinquency, impulsivity, gambling, involvement in housework, engagement in sports, and "Bem femininity" score (named after the psychologist Sandra Bem, who had constructed it). You can probably guess which of these received higher values in females and which in males, but you may be surprised to learn that 70 percent of the youth had both feminine and masculine characteristics.

Even more striking were the results from yet another dataset: a study of some 260 psychology students in a midwestern American university conducted by Bobbi Carothers, of Washington University, in St. Louis, and Harry Reis, of the University of Rochester.[3] It covered ten highly gender-stereotypical behaviors and activities: boxing, construction, playing golf, playing video games, scrapbooking, taking a bath, talking on the phone, watching porn, watching talk shows, and using cosmetics (the largest difference was in cosmetics: the chances of guessing the participant's sex on the basis of their use were above 90 percent). But when we looked at how these variables added up in individual students, less than 1 percent of the students had only "masculine" or only "feminine" characteristics; most—55 percent— were feminine-masculine mosaics.

If this mosaic of traits makes you question the very notions of masculinity and femininity, you are not alone. As I describe in the next chapter, in the past few decades, academics have proposed a radical revision of these concepts. This transformation is very much in line with my mosaic idea. And it echoes what we all know from our own experience: that having a good sense of humor does not imply you are also good at reading maps; being gentle does not mean that you are also good at expressing your emotions; and acting aggressively is no proof that you are good at math.

III

What's Wrong
with Gender

14.

From the Binary
to a Mosaic

The modern study of femininity and masculinity began in the 1930s, when Lewis Terman and Catharine Cox Miles published a 600-page volume, *Sex and Personality*. Starting from the belief that women and men "display characteristic sex differences in their behavior, and that these differences are so deepseated and pervasive as to lend distinctive character to the entire personality,"[1] they developed a questionnaire with 456 items on which men and women gave, on average, different answers. The items covered numerous subscales, including hobbies, interests, general knowledge, reading preferences, personality characteristics, and attitudes. In one instance, for example, respondents were asked to complete a sentence with a word that would make the statement true, choosing from several supplied, as in, "Things cooked in grease are: boiled, broiled, fried or roasted." Cooking was then still a strictly female domestic chore, so women were more likely than men to provide the correct answer, "fried," which was therefore scored as a female-typical response, whereas "boiled," "broiled," and "roasted" were scored as male-typical. For each person, the number of female-typical answers was subtracted from that of male-typical ones to assign the participant a so-called M-F score, which could range from −456, for highly feminine, to +456, for highly masculine. In reality, it ranged from −200 to +100 (average −70) for women, and from −100 to +200 (average +52) for men.[2]

For years afterward, psychologists continued to devise similar questionnaires, but they constantly ran into problems. The results obtained with the various questionnaires didn't fit the way people rated their own masculinity or femininity, or the way in which they were rated by others. Most troublesome were the poor correlations between the scores on different subscales; for example, people could come out masculine on "general knowledge" and feminine on "personality characteristics."

These observations should have led to the conclusion that humans possess a mosaic of gender characteristics, but scientists, just like the general public, had a hard time letting go of the masculine and feminine ideals.

Then, in the 1970s, two psychologists, Sandra Bem[3] (mentioned in the previous chapter in reference to her eponymous femininity score) and Janet Spence,[4] working independently, transformed the thinking about gender. Each of them developed a new questionnaire in which masculinity and femininity were measured separately. The Personal Attributes Questionnaire, for instance, developed by Spence and her colleagues, invited people to rate themselves on attributes highly valued in males, such as independence, self-confidence, standing up under pressure, and knowing the ways of the world, and attributes highly valued in females, such as being emotional, gentle, aware of the feelings of others, and liking children. Importantly, each person got a score on both the masculinity and femininity scales.

On the basis of scores obtained with these newly developed questionnaires, both Spence and Bem concluded that masculinity and femininity did not form a one-dimensional continuum but rather were two separate dimensions. If a person scored high on one scale, she or he was not necessarily low on the other.

This two-dimension approach, however, also ran into difficulties. Both Spence and Bem used narrow definitions: masculinity roughly corresponded to dominance, and femininity, to nurturance. But measurements on these scales did not fit in with other aspects of masculinity and femininity.

Finally, in the 1990s, Spence proposed that people

were "arrays" of masculine and feminine traits. She wrote in a 1993 paper that "men and women do not exhibit all of the attributes, interests, attitudes, roles and behaviors expected of their sex according to their society's descriptive and prescriptive stereotypes but only some of them. They may also display some of the characteristics and behaviors associated with the other sex."[5]

Still, Spence's concept didn't catch on. I hope it does now. My recent research findings have surely provided renewed support for it.

The recurrent observation that humans are mosaics of stereotypically female and male characteristics makes the notion of female and male natures as meaningless as that of female and male brains. Which of the endless personality mosaics that females display should be considered the female nature? And which of the many mosaics found in males can count as the male nature? Clearly, when we look at women or men as a group, there are average differences between the genders, and sometimes they even match the stereotypes. But average women and men are hard to find.

15.

———

Gender Illusions

So why, then, do women and men seem so distinct?

The answer lies in the very division of humans into two social categories: women and men. This division exerts a profound influence on the way we act and how we view the actions of others. Behavior, and the way we perceive it, reflects not only people's mosaics of abilities, qualities, and preferences but also the roles they play in society, the situations they find themselves in, their status, and their own and other people's expectations. All these are different for men and women in our society— and they create the illusion that humans belong to two distinct types.

We typically encounter women and men in different situations, which bring out different aspects of their personalities, but we tend to attribute the differences in their behavior to sex rather than context. For example, we are likely to see more men than women in a business meeting, and more women than men on children's playgrounds. These two contexts call for different behaviors, but we attribute the differences to people's sex: we conclude that men are assertive and women are nurturing.

In my lectures, I sometimes show a photo of myself from several years ago in which my three young kids are all over me, tousling my hair. I invite the audience to imagine what they'd think of me if they first met me in this situation. They would probably decide that I'm much nicer and warmer than the lecturer standing in front of them; on the other hand, if they first met me as a lecturer, they'd be more likely to think that I'm a scholarly type. Evidently when I switch roles, my abilities do not change but my behavior does.

Similarly, much has been written about differences in the way men and women speak, but on closer inspection, these are often revealed as differences of status, not sex. Both men and women of a higher status tend to assume a style of speech considered masculine (for example, avoiding eye contact and interrupting the other person),[1] whereas lower-status men and women assume a style viewed as feminine (for example, smiling for no reason).[2] When my students—males and females alike—send me an email, they start with "Dear Professor Joel," politely spelling out the purpose of their address and ending it with "thanks" and "best regards"; my reply to them is often brief and matter-of-fact. But when I send an email to the president of my university, I'm just as courteous as my students are in addressing me.

These examples show how easy it is to mistake adjustments to context for gender differences. But often men and women do act differently in the same context—not because they differ, biologically or psychologically, but merely by virtue of belonging to one gender or the other. Our expectations from ourselves as women or as men, as well as the expectations of others, "gender" our behavior, even when we are not consciously aware of those expectations. For instance, studies have shown that women tend, on average, to take less credit for success in public than in private—because modesty is considered a feminine virtue; men, in contrast, rate their achievements similarly in both settings.[3] On the other hand, studies show that men, as opposed to women, are more likely to stick up for their beliefs in public than in private— because men aren't supposed to be pushed around.[4]

Our gendered expectations of ourselves also affect our level of performance. One example is what's called a stereotype threat—dread that a poor performance will confirm a stereotype about our group's intellectual inferiority. Studies show, for instance, that the ability to perform mental rotation can be undermined by the stereotype threat. In one study, when students were asked questions about their gender before taking a mental rotation test, female students did significantly worse, on average, than male students. But the gap between the sexes became much smaller when the students were asked instead to contemplate the fact that they were enrolled in an elite college.[5]

The division of humans into two genders affects not only our own behavior but also the way we judge the behavior of other people. We view them through gender schemas, or stereotypes—that is, our beliefs about women and men.

Studies in social psychology reveal that once a gender stereotype has taken root, it is extremely hard to change. We tend to perceive and remember details that match the stereotype and ignore those that do not. We believe information that matches the stereotype much more than information that does not. Finally, when we encounter a characteristic that matches the stereotype, we attribute it to sex, and when it doesn't, we write it off as an exception or attribute it to circumstances or to a person's individuality.

For example, one often hears that boys "naturally" like to play with blocks and trucks, and girls "naturally" choose dolls. So if a boy likes blocks, his parents say, "He's such a boy!" But if he also likes dolls, they say, "My Danny is so special, he loves babies." If a girl plays with dolls, her parents say, "She's such a girl!" But if she loves to play football, they say, "Ruthie is just like her father."

Because most of us possess a mosaic of gender characteristics, it is easy to see in each person just what we expect to see: a man or a woman, a boy or a girl—as did my friend Lisa with her three daughters. Moreover, by ignoring traits that don't match the stereotype or attributing them to other factors, we manage to preserve the stereotype even in the face of conflicting evidence. People might say about a sentimental man, "He is probably a poet," or about a man who is a bad driver, "Well, he is busy with his cell phone," thus preserving the stereotype that men are not sentimental and are good drivers.

We sometimes even deal with counter-stereotypical behaviors not by changing the stereotype but by reinforcing it. The word for a girl who likes to climb trees is "tom*boy*." Thus an encounter with an adventurous girl reinforces the stereotype that it is boys, rather than girls, who tend to be adventurous.

So if we take a male and a female with an identical mosaic of characteristics, not only will they behave differently because one is labeled a "man" and the other a "woman," but we will perceive them differently for the same reason: seeing him as more man-like than he actually is, and her as more woman-like.

Plenty of external cues ensure that we easily identify each person's gender from the moment of birth. We dress a baby girl in pink and a baby boy in blue[6] and continue to mark females and males differently throughout their lives by clothes, cosmetics, jewelry, and hairstyle—which ensure that even on first encounter, instead of seeing the unique mosaic of a human being, we see a woman or a man.

What also interferes with acknowledging the mosaic is the implicit assumption that humans are internally consistent—that is, consistently masculine or consistently feminine. When we encounter a person who combines extremely feminine and extremely masculine traits so prominently that we cannot disregard the gender-"inappropriate" ones, a common response is to be skeptical.

When people learn that a football player loves poetry, they are incredulous: "But he is such a man!" When they hear that a woman loves watching boxing matches, they exclaim in disbelief: "But she's so feminine!" One of my kids is a big fan of soccer, which in Israel is the men's ball game; he roots for Barcelona and for many years was busy doing the math to figure out whether he and Lionel Messi could one day meet on the field. His soccer teammates find it hard to believe that he also plays the piano and loves babies.

Or take the example of Ayelet Shaked, a hard-line politician, who at the time of this writing served as Israel's minister of justice. She's also a strikingly beautiful woman who lets her lush hair loose and occasionally wears multicolored outfits. Throughout Shaked's term in office, there's been much hullabaloo in the media not only over her tough, controversial policies—for instance, on redefining the authority of the Supreme Court of Israel—but also over her looks.[7] Her choice to magnify her femininity seemed to baffle commentators on both sides of the political spectrum. But all such discussions would have been moot if only reporters acknowledged that Shaked, like the vast majority of us, possesses a unique mosaic of masculine and feminine traits, rather than viewing her through the prism of gender schemas.

The belief that gender characteristics go together can, in certain situations, make us avoid or conceal behaviors considered highly appropriate to our gender, so that we are not judged unqualified for tasks traditionally belonging to the other gender. It's something we may instinctively do from a young age. As a kid, I used to love playing soccer and basketball with boys from the neighborhood, so I instructed my younger sister not to tell them that at home I played dolls with her—I was worried they'd shut me out if they knew. Female politicians often cut their hair short and wear solid-color suits. Similarly, women who work in other male-dominated fields often play their feminine identity down. Female colleagues of mine in "masculine" disciplines such as computer science tend to avoid wearing lipstick and high heels in order to be taken seriously as scientists. In "male territory," displaying feminine attributes might mean an uphill battle—in real life, and in books and movies. In the movie *Legally Blonde,* the heroine goes through hell and high water before managing to prove that she can be a blond bombshell *and* a brilliant lawyer. She struggles for her right to put her human mosaic on display.

The myth of the existence of two genders is just that: a myth. It does not stand up to the test of modern science. Myths are stories that explain why the world, natural or social, is the way it is, and why it could not be different. We easily identify a myth when it belongs to a different culture. But we treat the myths of our own culture as truth.

Myths that explain a gendered world have existed throughout history: that Adam was created before Eve; that men are smarter than women because they have larger skulls; that women have a female brain and men have a male brain. All these myths justify the existence of gender as a social system in which humans are treated differently and have different access to power—depending on the form of their genitals.

The fact that gender is a myth doesn't mean it doesn't exist. It surely does—not as an intrinsic set of qualities, but as a social system that attributes meaning to sex, assigning different roles, status, and power to males and females. This system exerts an overriding influence on our lives, imposing a binary divide on a population of human mosaics.

16.

Binary
Brainwashing

When I submitted my very first scientific manuscript for publication in the mid-1990s—together with Ina Weiner, my PhD adviser—we used our initials instead of our full first names. It was a provocative paper; we challenged the then dominant view about the way the frontal lobes are connected with the basal ganglia, a group of deep-seated neuronal clusters underneath the cortex. We were concerned that if the reviewers knew we were women, our chances of publishing this manuscript would be nil. I don't know if the initials helped, but the paper did get published, in the journal *Neuroscience,*[1] and became one of several that changed the dogma.

Call us paranoid, but I later learned that at least we had been in good company. It turns out that at around the time we'd submitted our controversial paper, the author Joanne Rowling had adopted the initials "J K" as a pen name at the request of her publisher, who feared boys might not read a book written by a woman.

In case you think the latent bias against women has since vanished, a study published in 2012 in the *Proceedings of the National Academy of Sciences, USA*, showed to what extent it hadn't. Yale University researchers obtained feedback from 127 professors in biology, chemistry, and physics from several universities across the United States who had been asked to evaluate the credentials of a fictitious student supposedly applying for the position of lab manager.[2] All the professors, who were unaware they were taking part in an experiment, received the same application. There was only one difference: in about half of the cases, the applicant's first name was stated as John; in the other half, it was Jennifer. You may have guessed who fared better. The professors rated John as significantly more competent and hirable than Jennifer and suggested granting him a higher starting salary, of about $30,000 on average, compared with about $26,500 for Jennifer.

Interestingly, female faculty members were just as likely as their male colleagues to favor John—in line with other studies, showing that women are often no less gender-biased than men. And it wasn't a matter of hostility toward women, just the opposite: the professors reported liking Jennifer more than John. Study authors suggest the bias was probably "unintentional, generated from widespread cultural stereotypes rather than a conscious intention to harm women."

Even if unintentional, such biases produce their toxic results. They affect every aspect of a woman's career—from her choice of studies to her chances of getting a job, the salary she will earn, and her prospects of being promoted.

It's not always easy to spot how and when gender channeling takes root, but people, it turns out, already treat babies differently by gender in the crib. In one study, researchers at City University of New York conducted what they called a "baby X" experiment: they dressed a three-month-old baby in a gender-neutral yellow jumpsuit and observed how adult volunteers interacted with the infant.[3] The volunteers were told the purpose was to study infants' responses to strangers, but the real object was to reveal their own gender biases, if any. Indeed, those who were told that the baby was a girl were twice as likely to offer the baby a doll out of a choice of toys than those who were told the baby was a boy. This study and similar ones show that the different treatment female and male babies receive is dictated by their gender labels and not by the infants' supposedly differing behaviors.

Such gender channeling can easily turn detrimental. In one study, published in the *Journal of Public*

Economics, Victor Lavy, of the Hebrew University of Jerusalem, and Edith Sand, of the Bank of Israel, compared the grades boys and girls received from their elementary school teachers, most of them women, with the scores the same children received in a national exam by examiners who were unaware of the pupils' sex.[4] The comparison revealed that some of the teachers systematically favored one gender over the other. Some were biased against boys, but about twice as many were biased against girls. On average, the girls taught by these teachers received a lower grade in class than on the national exam, whereas boys taught by these teachers received a higher grade in class than they did on the national exam. Most troubling was the finding that this teachers' bias had long-lasting consequences. Girls who in fifth grade had a teacher biased against female students were less likely than other girls to take an advanced-level math course in high school and obtained lower grades in high-school matriculation exams, whereas boys from such classes fared better than did boys taught by nonbiased teachers.

So a pair of initials or an anonymous exam can free our perception from viewing others through gender schemas. Unfortunately, in most situations, we cannot hide behind initials or blind ourselves to the gender of another person. Nor can we apply these tactics to the way we perceive ourselves.

I find it particularly disturbing that self-stereotyping

by gender starts early in life. Researchers Andrei Cimpian, of New York University, and Sarah-Jane Leslie, of Princeton University, set out with colleagues to test the impact on young children of a long-held myth, that genius is all about men—white men, to be precise. More than a century ago, back in 1909, this belief was captured, for example, by Nobel Prize–winning biologist Elie Metchnikoff, who told the *New York Times* in an interview: "Genius, I believe, is a masculine quality, just as a beard is, for instance, or as strong muscles are."[5] Sadly, as Cimpian, Leslie, and Lin Bian reported in 2017 in *Science,* this historic myth persists in the twenty-first century.[6] The researchers called the modern version of this myth "the brilliance trap." In an article in *Scientific American,* Cimpian and Leslie wrote about their study: "We asked hundreds of five-, six-, and seven-year-old boys and girls many questions that measured whether they associated being 'really, really smart' (our child-friendly translation of 'brilliant') with their gender. The results…were consistent with the literature on the early acquisition of gender stereotypes yet were still shocking to us."[7]

The researchers found no difference in self-assessment between male and female five-year-olds, but just around that age, the stereotype had somehow begun to creep in. By age six, girls were less likely than boys to think that members of their gender are "really, really smart." "The more a child associated brilliance with the opposite gender, the less interested he or she was in playing our games for 'really, really smart children,'" Cimpian and Leslie wrote. "This evidence suggests an early link between stereotypes about brilliance and children's aspirations. Over the rest of childhood development, this link may funnel many capable girls away from disciplines that our society perceives as being primarily for brilliant people."

Indeed, in a separate study, Cimpian, Leslie, and colleagues found that fewer women and African Americans were awarded PhDs in academic disciplines considered to be heavily dependent on brilliance, such as philosophy, physics, and certain subfields of math—compared with areas that are less associated with genius, such as psychology and molecular biology.[8] The implications of this study were best summed up in the reading line at the head of the *Scientific American* article: "How a misplaced emphasis on genius subtly discourages women and African-Americans from certain academic fields."

There are, of course, much more severe effects of the binary divide. In some traditional societies, women are not allowed to own property or inherit it from their parents, to open a bank account, to drive or leave the house without being accompanied by a male family member. In certain societies, the restrictions can even extend to murder; a man may kill a woman if he decides that she'd brought shame upon her family. And in Western societies, women are all too often subjected to sexual harassment and assault, as we've been amply reminded by the #MeToo movement.

Men, too, face substantial risks forced upon them by the gender binary. Its consequences for girls and women are more readily recognized, but boys and men can be harmed by virtue of making up the gender system's dominant group.[9]

For instance, after collision with an iceberg has wrecked the *Titanic* in the eponymous film epic, the young lovers, played by Leonardo DiCaprio and Kate Winslet, having jumped into the ocean, come upon a wooden panel that can hold only one person. Which one of them will get to climb onto the lifesaving board? So merciless are the dictates of modern-day patriarchy that this question doesn't even cross the viewer's mind. Of course, it's the man who will help the woman onto the panel, in a laudable surge of chivalry, himself remaining to face certain death in the icy waters.

Even when they don't find themselves in a shipwreck, men often get a raw deal by virtue of belonging to the group empowered by the patriarchal order. It is mostly men who die in droves in wars, are injured in work-related accidents, and feel compelled to become providers, often at the expense of following their hearts to a career in the arts or other non-bread-winning fields. All these follow directly from being part of the dominant group. Being a member of this group means not only having and exerting power—it also means helping those who are not supposed to have power and suffering from the stresses imposed by the demands of dominance.

Just as with girls, the harmful effects of the gender binary on boys start early. One of the most worrying outcomes is in the realm of emotions.

In the first two years of life, baby girls and boys do not differ, on average, in their display of feelings, Lise Eliot, of Rosalind Franklin University of Medicine and Science, concluded in her book *Pink Brain, Blue Brain,* in which she reviewed extensive research on this topic.[10] Baby boys laugh, cry, and become upset, happy, annoyed, or shy, just as baby girls do. Of course, there are large differences between individual babies but, studies show, not between boys and girls as a group. Yet people surrounding the babies—parents, grandparents, older siblings, and others—treat baby girls and boys differently. They tend to speak more to baby girls than to baby boys, in general and about emotions in particular—except when it comes to anger, a topic they discuss more with boys than with girls. They also treat emotional displays of babies differently.

When a baby girl cries, she is more likely to be comforted by others than a baby boy is. He, on the other hand, is more likely to be ignored and, at an older age, even to be reprimanded—"man up," "boys don't cry," "stop crying like a baby/girl." When a baby girl is angry, she is more likely than a baby boy to be ignored or reprimanded—"be nice," "act like a lady." Eliot cites studies showing that later in life, starting around two to three years of age, sex differences start to emerge. Girls express more emotions than boys in speech and in facial expression, except anger—which boys express more than girls. As I explained in chapter 4, we cannot say for sure that these sex differences in the display of emotions are caused by the different treatment the children had received as babies, but the parallels are striking.

It's most unlikely, by the way, that the brain circuits underlying emotions are deficient in boys and men, restricting their expression. Anyone who's ever watched the World Cup can attest to having witnessed a full gamut of emotions on the faces of the players (all men) and of the men in the audience. But outside the soccer field, boys and men commonly shy away from freely expressing the full range of human feelings.

The differences in how we treat girls and boys are typically small, but coming from many people over one's lifetime, they end up harming both sexes. By placing kids into emotional straitjackets of gender, we raise power-disabled girls and emotion-disabled boys.

Surely, not knowing how to control your anger can get you into serious trouble, but not knowing how to *express* anger when appropriate—for example, when you discover you've been wronged—can also be detrimental. By teaching girls to suppress their anger, we are restricting their assertiveness, a trait needed later in life for negotiating a salary or promotion, or otherwise getting what they deserve.

Men pay a price for being expected to always be strong

and discouraged from freely expressing their emotions. The 2015 documentary *The Mask You Live In* explores the impact of what it calls "the three most destructive words" boys hear when growing up: "Be a man!" In this film, one interviewee after another describes how the imperative to hide their feelings, having anger as the only acceptable emotional outlet, robs them of warmth, care, and closeness.

The constant need to prove their masculinity can also adversely affect men's mental health. A study conducted at Indiana University–Bloomington found that men who felt obliged to conform to masculine norms suffered more from depression and substance abuse and were less likely to seek psychological help than men who didn't conform to these norms.[11] Three aspects of idealized masculinity were shown to be particularly harmful for men: the need to be self-reliant, to act like a playboy, and to have power over women. (Even though the effects of these perceptions of masculinity on women were not part of the study, there is no doubt women are also harmed by the latter two.)

I'm not suggesting a contest to determine who suffers more from the gender system, women or men, not to mention the sufferings of people who do not fit into this binary system. The fact that we all suffer from the gender binary seems to me a good enough reason for doing away with it.

IV

Toward a World
Without Gender

17.

How to Deal with Gender Myths

In July 2017, Google engineer James Damore wrote an internal company memo[1] that generated a global media storm. He stated that Google's gender diversity policies were discriminatory and unfair to men. The reason there are few women among software engineers, he argued, is that women don't naturally possess the traits needed for this job. The "abilities of men and women differ in part due to biological causes and...these differences may explain why we don't see equal representation of women in tech and leadership," he wrote. He went on to list qualities and abilities—for example, cooperativeness and interest in people rather than things—on which women and men differ on average, claiming that no policy can achieve diversity without acknowledging "population level differences" in the distribution of these traits.

Damore was fired from Google, but not for getting the science wrong. His memo, despite some inaccuracies, draws on bona fide academic journals, and it represents beliefs about women and men that can, unfortunately, be found on the pages of reputable academic publications. Some of the sex differences on his list had been demonstrated in large studies or meta-analyses. For example, compared to men, women show, on average, a higher interest in people and a lower interest in things,[2] score higher on the anxiety component of neuroticism and lower on the assertiveness component of extraversion,[3] and are more concerned with work-life balance. Other qualities Damore listed, not necessarily supported by studies, are nonetheless deeply rooted in common beliefs—for example, that men have a higher drive for status and that women are more sociable and cooperative.

The main problem with these "women are like *this*, and men are like *that*" stories is not so much the evidence itself but its interpretation. When people put together such stories, they tend to cherry-pick pieces of evidence to explain why the world is the way it is—for example, why there are few women in tech and leadership—instead of taking an unbiased look at *all* the evidence to try to figure out what is going on.

Let's start with the fact that any job requires a combination of qualities. When people make up lists of qualities needed, for example, for being an engineer at Google, and try to explain that there are more men than women in this field because men score higher on these qualities, they pick traits that are, on average, more prevalent in men—and those that, to their mind, support their claim. Qualities that may contradict the claim—some prevalent in men, others in women—are absent from their lists. Thus, Damore's list of desirable qualities includes a drive for status. Yet in an industry that's as reliant on teamwork as high tech, a high drive for status may interfere with success. High cooperativeness should logically be a plus in this industry, but it doesn't appear among Damore's desirable traits; on the contrary, he cites it among the qualities that may be hindering women's success.

Researchers sometimes use a similarly selective approach when seeking to explain social phenomena in animals. For years, they had attributed the fact that males dominate females in many mammalian species to the males' being larger and more aggressive. When females were found to dominate males—as, for example, among spotted hyenas—this was attributed to their being, on average, larger and more aggressive than males. But in 2018, researchers from the Leibniz Institute for Zoo and Wildlife Research reported in *Nature Ecology & Evolution* that in fact, among spotted hyenas, it's the level of social support that an animal has, not its body size or aggression, that determines whether it will win or lose in one-on-one interactions and, consequently, defines its rank in the hierarchy.[4] And in this species, it was often females who had more social support than males. The authors note that in other species, too, including primates—and humans are primates—"individuals with greater social support may be more assertive and more likely to win an encounter, even when their coalition partners are absent or do not intervene."

When compiling lists of attributes needed for success in a particular occupation, people also ignore the fact that the same job can usually be done well by people with significantly different sets of qualities. This is surely true for members of a team. Managers, too, are not all cut from the same cloth. In trying to motivate the team, say, to spend Friday night in the office in order to meet a deadline, one manager might rely on her charisma, another on his ability to create a family-like ambience, while yet another, on his or her powers of intimidation.

Moreover, if you've made it this far in this book, I hope to have convinced you that people possess a mosaic of qualities, which don't add up consistently in any one person; different people have different mosaics. So when we're drawing up a list of desirable qualities, even if all are "masculine," we should not assume that if someone is high on one "masculine" quality, he or she will also be high on others. A man may be career-oriented but not necessarily assertive, or—as reported in the *Journal of Counseling Psychology,* in a meta-analysis covering studies of over a million participants—he may be interested in things *and* in people.[5] If people didn't possess different mosaics of traits, we'd need to assume that every candidate with suitable masculine qualities must also be physically aggressive—hardly something you'd want in prospective hires for most jobs—because physical aggression, on average, is more common in men than in women.[6]

Selective gathering of evidence is not limited to lists of required qualities. Even the way in which these traits are measured is often biased to match gender stereotypes. One occasionally hears claims, for example, that investment banking and start-up entrepreneurship are male-dominated because men are more prone to risk-taking than women. But as Cordelia Fine, of the University of Melbourne, points out in her book *Testosterone Rex,* the "evidence" cited in support of this claim follows an inverted logic. Because men are considered more risk-taking than women, questionnaires assessing risk-taking include activities that are more likely to be undertaken by men than women—for example, skydiving. Fine suggests that "the reported gender gap in risk taking would almost certainly narrow if researchers' questionnaires started to include more items like *How likely is it that you would bake an impressive but difficult soufflé for an important dinner party, risk misogynist backlash by writing a feminist opinion piece or train for a lucrative career in which there's a high probability of sex-based discrimination and harassment?*"[7]

Even when gender differences on single traits are measured in an unbiased manner, the gender myths— "men are like *this,* women are like *that*"—distort the evidence by presenting these differences as binary. But, as we've seen time and again, there's a great deal of overlap between the sexes on any individual measure. For instance, as mentioned above, women score higher than men, on average, on the anxiety component of neuroticism, but according to a statistical analysis of numerous studies, 38 percent of men score higher than the average woman on this trait.[8]

Or take the gender difference in mental rotation of three-dimensional geometrical shapes, an ability often considered relevant to engineering (I discussed sex comparisons of neural mechanisms behind mental rotation in chapter 8). Men, on average, get higher scores than women on this task. In one large study by researchers from the University of Guelph, which included over 130,000 men and 120,000 women, the average mental rotation scores were 3.16 for men and 2.15 for women— out of a total of 6 points.[9] But about 30 percent of women scored higher than the average man. Assuming that performance in this task is indeed important for an engineer, and given this overlap, as well as the relatively low average scores for both sexes, what would be the best strategy to adopt when hiring an engineer, at Google or elsewhere—to automatically assume that male applicants are exceptionally good at this task, or to give all applicants a test of mental rotation and choose the best ones?

Finally, there's the unfounded assumption that observed gender differences are "natural." These differences are regarded as "inborn" or preprogrammed, which, in turn, implies that there is nothing we can or should do about them. But—again, as we've seen—it's usually impossible to conclude that a difference between women and men is a direct consequence of sex—that is, of sex-related genes and hormones. On the other hand, there's plenty of evidence that gender contributes to observed differences between women and men, affecting how various abilities develop, how they manifest themselves in behavior, and how they are perceived (as happened to the fictional John and Jennifer described in the previous chapter).

But the gender system does more than distort our abilities and perceptions. It can make women feel unwelcome in male-dominated fields, causing them to quit or preventing them from entering such fields in the first place. A report released in June 2018 by the National Academies of Sciences, Engineering, and Medicine describes pervasive "gender harassment" of female students—behaviors that belittle women and make them feel they don't belong.[10] In a large survey conducted by the University of Texas, 50 percent of female medical students and 25 percent of female engineering students reported having been the butt of demeaning jokes or comments that women are not smart enough to succeed in science. Such a hostile atmosphere can interfere with a woman's performance and may even push her out.

Sexual harassment is yet another obstacle to women, especially when men largely have the clout in their professional field. Women working in Silicon Valley report being commonly propositioned, touched without their consent, and being the target of sexist comments, according to a story by Katie Benner in the *New York Times*. "The women's experiences help explain why the venture capital and start-up ecosystem—which underpins the tech industry and has spawned companies such as Google, Facebook and Amazon—has been so lopsided in terms of gender," wrote Benner, the paper's Justice Department reporter.[11]

Taking into account all these and other obstacles faced by women in their careers in a gendered world, it seems to me that removing the wrongs of the binary gender system in every way we can would be much more productive in eliminating discrimination than making up selective lists of dubiously "natural" traits to explain existing gender gaps. After we do away with all gender barriers, I have no idea whether it will be men or women who will make up the majority in this or that field or occupation, for example, among engineers at Google. But I couldn't care less.

What I do care about is creating an environment in which all people can choose their path according to their individual mosaic of attitudes, preferences, and traits. Starting in chapter 19, I will discuss what we can do to bring such an environment closer to reality. But first let me share a few facts and thoughts about emerging cracks in the gender binary.

18.

Gender Blending

When I was eighteen, a male friend of mine who'd just been accepted to the pilot training course in the air force told me that women couldn't be pilots because they weren't good at doing two things simultaneously. When I was thirty-four, the father of my kids told me he couldn't change a diaper while talking on the phone because men couldn't do two things simultaneously.

In case you wonder which one of them was right, the findings on multitasking follow the same pattern we've seen in many other comparisons between women and men. Some studies report that women are better, on average, at this cognitive ability, others say it's men, and in all studies, the average difference between the two genders is small, with a great deal of overlap in the scores of individual women and men.[1] This pattern amply allows for real or imagined differences between the sexes to be exploited to serve almost any agenda. But in any event, it's the person's own score, not the average one for either sex, that determines her or his ability to become a pilot or to change diapers while talking on the phone.

To the many arguments over whether women and men are the same or different, I say: they are neither. We are all different. Each one of us is a unique mosaic of features. The fact that some of these features are more common in females and others more common in males is irrelevant to who we are. Or at least I believe this is how it should be.

I therefore find it extremely encouraging that in Western societies, the binary gender system seems to be crumbling. In everything related to gender, diversity is increasingly entering our awareness—and the mainstream, despite an accompanying backlash. People with unconventional gender behaviors, once marginalized, are now prominently featured in movies, on TV, and on stage. When recording artist Thomas Neuwirth faced prejudice for being gay while growing up in Austria in the 1990s, he probably couldn't imagine that some two decades later, he would become a global superstar after adopting a bearded-female stage persona as Conchita Wurst and winning the 2014 Eurovision Song Contest.

Diversity is apparent even in gender identity, a core psychological parameter on which females and males greatly differ, on average. When I signed up at Facebook, I recall choosing between "male" and "female" in the gender section. But in the past few years, Facebook has been offering its users new "custom" gender options for the personal profile; the list numbered seventy-one options last time I checked, including "pangender" and "gender nonconforming."

Even state and national governments have started to recognize nontraditional forms of gender expression. While I worked on this book, their lists kept growing. In June 2017, the District of Columbia began to offer the gender-neutral choice of "X" on driver's licenses and identification cards; in August 2017, Canada joined the list of at least ten countries—among them Germany, India, and New Zealand—that allow citizens to indicate their gender as "unspecified" on passports and other government documents; in January 2019, New York City started offering gender-neutral birth certificates, joining California, Oregon, and Washington state, which had begun to offer this option a year earlier.

Diversity in gender identity is apparent even in people who regard their own gender in conventional ways—as a woman, for people assigned the female sex at birth, and as a man, for people assigned the male sex at birth. Research shows that these so-called cisgender individuals are far less binary in their gender identities than might be expected. In one study of this phenomenon, my colleagues and I drew inspiration from reports that some transgender individuals experience themselves as belonging to both genders or to neither.[2] We asked 2,155 cisgender people to grade themselves in an anonymous questionnaire: how often they thought of themselves as a man and how often they thought of themselves as a woman, on a scale of 0 (never) to 4 (always). We didn't try to cover all types of gender experience, only the two most common ones—man and woman. Nor did we try to clarify what meaning each person invested in the terms "man" and "woman"; we only asked the questions and mapped out the responses. Predictably, as a group, males felt more "like a man" than females did, and females felt more "like a woman" than males did. But the startling finding, reported in *Psychology & Sexuality* in 2013, was that about 35 percent of participants perceived themselves as both a man and a woman.[3] And these were people who self-labeled as "man" or "woman," not as "transgender" or "other."

We followed up with a larger, international study, published in 2018 in the *Archives of Sexual Behavior,* in which we questioned 4,759 cisgender men and women from several English-speaking countries about their gender identity.[4] Here again, 38 percent of the participants said they felt to some degree like the other gender. A similar percentage of participants also revealed they sometimes wished they'd belonged to the other gender, or wished they had the body of the other sex—sentiments typically associated with transgender individuals or those with nonbinary gender identities. The finding that so many cisgender people experienced such feelings supports the stand of the World Professional Association for Transgender Health: that gender variance not be treated as a pathology.[5]

Transgender individuals themselves, as mentioned, vary widely in their gender identity, and this was also evident in a study we published in 2018 in the *Journal of Sex Research*.[6] Many of the 406 transgender participants said they felt like both a woman and a man, or like neither. Somewhat surprisingly, feeling like the "other" gender (compared to the gender they were assigned at birth) was not always associated with a strong wish to have the body of the sex that typically goes with that gender.

I say "somewhat surprisingly," because current cultural portrayals of transgender people stress this wish as a central aspect of the transgender experience. But our results reinforce the claims made by Kay Siebler, of Missouri Western State University, and others, that the hormonal or surgical interventions sought by some transgender people may sometimes serve social expectations rather than their own need for aligning their physical appearance with their psychological experience of gender.[7]

Broadening the norms for how gender identities can be expressed will relieve the pressure not only from transgender but also from cisgender individuals: it will capture the experience of many that gender identity cannot be compressed into the man-woman binary. This will surely alleviate the stress that comes with atypical gender expressions, and may also reduce the pressure that some transgender and gender-nonbinary individuals feel to undergo body modifications.

A blending of gender identities can already occur in childhood, as documented by researchers from Arizona State University and New York University.[8] In a study reported in *Child Development* in 2017, the researchers asked 467 boys and girls in first, third, and fifth grades whether they perceived themselves as similar to their own or to the other gender. Around 30 percent of the children said they felt similar to both boys and girls, whereas about 17 percent claimed they felt similar to neither.

What makes all these findings remarkable is that we are repeatedly told that gender is "either-or": if you are female, you'll grow up to be a woman; if you are male, you'll grow up to be a man. Yet even in a world that labels us in terms of one gender or the other, lots of girls and boys, women and men, believe themselves to be something of both, or neither.

Research conducted in the past few decades, including my own studies, has shown that humans vary greatly in their psychological characteristics and do not fit into an either-or, woman-or-man framing of human nature.

I hope that in a not-too-distant future, this idea will

be taken for granted; that gender studies will be a history course; and that when the topic of gender comes up, children will need to ask their parents (or grandparents) to explain why on earth someone had once thought people had to be grouped by their genitals.

19.

Gender-Free
Education

When I was in second grade, children in my school were offered a choice of extracurricular classes that were open to girls and boys, but one crafts class was for boys only. My mother couldn't answer my question—Why weren't girls allowed to take this class?—and suggested I ask the teacher. The teacher didn't know the answer, either, and suggested I talk to the school principal. That was how, at age seven, I found myself in the principal's office, on my very first feminist mission.

The principal didn't have an answer to my question, but soon after that discussion, she announced that the crafts class would be open to girls as well. Several weeks later, at a parents' meeting, she asked my parents if I was enjoying the crafts class, and was surprised to hear I hadn't taken it. My mother explained: "Daphna didn't know if she wanted to take this class, she just wanted to make sure she could." (For the record, I did take the class the following semester, and to the best of my memory, I enjoyed working with wood, the main activity it offered.)

I believe children should have access to an array of options, regardless of the form of their genitals. If a kindergarten teacher invites girls to listen to a story and tells boys to go play with a ball in the meantime, he or she restricts the options of both sexes. In doing so, teachers reveal their own, probably implicit, assumption that there are two types of kids: those who like to play ball but not listen to stories, and those who like stories but not ball games. But as we've seen throughout this book, human traits are not divided neatly into two sets.

For one, even if girls and boys differ, on average, in the degree to which they possess a particular trait, or in their preference for a certain activity, there is always overlap: some children will have the trait that's more commonly observed in the other sex, or will prefer activity that's typically reserved for the other sex. There is no reason to limit their possibilities just because they want to do something that's not so common for children of their sex—just as you wouldn't keep your son from playing chess just because most boys prefer soccer.

If the teacher offered each of the two activities separately, it is indeed possible that more girls than boys would take up the invitation to listen to stories, and more boys than girls would go for ball games. There will also be children who will want to engage in *both* activities, while yet others might choose neither. That's the mosaic concept in its most basic form—take only two preferences, and we already get four "types" of children, not just two. The more activities we consider, the more "types" we'd get. Forcing the richness of the children's choices into just two types is bad enough, and doing so according to the form of their genitals adds insult to injury.

Studies show that gender stereotyping reduces the chances that children will choose to engage in activities labeled as "inappropriate" for their gender, enjoy them, and perform well in them.[1] This alone is reason enough to avoid labeling activities as being "for girls" and "for boys."

As parents or as educators, I believe we should encourage children to act and express themselves irrespective of what society deems appropriate for girls or boys: playing ball *and* reading books, standing their ground *and* showing empathy, working hard to achieve their aims *and* being able to voice sadness or frustration. Moreover, we, as a society, should open up all fields to everyone and encourage people to develop the full range of human virtues, without connection to sex, just as we wouldn't limit access to a certain area based on skin color.

For every behavior, we must make ethical decisions on whether it is appropriate for humans. We should not concern ourselves, in this respect, with whether a behavior stems from nature or nurture, nor with whether boys or girls are "naturally" gifted for this or that skill. If violence is considered detrimental, let's restrain it, even if it is "natural" for humans to be violent sometimes. If we think math, sports, assertiveness, empathy, and the ability to express emotions are important, all children, female or male, should be encouraged to master those; if a child has difficulty with any of these—regardless of whether the difficulty stems from genes, hormones, parental treatment, or culture—the child should get help in order to reach a certain level, as we would have done with difficulty in reading.

The earlier we start, the better. In one welcome development, which I hope becomes a trend, a number of stores in different countries have let it be known that they've stopped labeling toys as being "for boys" and "for girls." For example, the American retail giant Target announced in 2015 that in response to customers' concerns about unnecessary gender-based signs, it would ditch gender labels on toys. A campaign, Let Toys Be Toys, that petitions retailers in the United Kingdom to "please sort toys by theme or function, rather than by gender, and let the children decide which toys they enjoy best," reports a decrease in the number of gender labels in stores and on websites. It does note, however, that in toy catalogs, "children's play is still represented in very stereotypical ways, with boys four times as likely to be shown playing with cars, and girls twelve times as likely to be shown with baby dolls."

Starting early also means ridding the education system of gender long before first grade. Sweden, consistently ranked by the World Economic Forum as the fourth most gender-equal society in the world, has been a leader in gender-free education for young children. Its national curriculum requires that preschools "counteract traditional gender roles and gender patterns," for instance, by avoiding gender labels for toys and activities, using the recently adopted gender-neutral pronoun *hen* instead of *han* (he) and *hon* (she), and addressing children as "friends" rather than "boys and girls."[2] Some of the preschools have also started coaching children in activities associated with the other sex, by putting boys in charge of the play kitchen and increasing assertiveness in girls by having them practice shouting "No!"[3]

One of the most difficult things to change, it turned out, has been the gendered attitudes of teachers toward children. When teachers watched video of themselves interacting with children, they were shocked to see how differently they treated boys and girls. Many used more complex sentences and richer vocabulary with girls than with boys; one teacher realized she was helping the boys bundle up before going outside into the cold but expected the girls to dress themselves. "It was hard at first to see patterns," this teacher told a newspaper reporter. "We saw more and more, and we were horrified at what we saw."[4]

It's too early to know the long-term consequences of gender-free preschools, especially since the children still live in a gendered world. But a small study, published in 2017 by Swedish and American psychologists, found that, compared to children attending traditional kindergartens, those in gender-neutral kindergartens were more willing to play with unfamiliar peers of the opposite sex and made fewer gender-stereotypical assumptions about them.[5] The potential benefits of a gender-free environment—including fewer discipline problems in the classroom and improved self-confidence in girls— also became manifest in *No More Boys and Girls: Can Our Kids Go Gender Free?* This two-part BBC documentary followed a bold social experiment in which Dr. Javid Abdelmoneim attempted to free a group of seven-year-olds from the gendered treatments girls and boys commonly receive, such as having the girls being called "love," and the boys "mate," and being offered gendered toys and activities.[6]

We shouldn't expect gender-free education to erase average differences between the sexes. A gender-free world doesn't mean such differences wouldn't exist— they may or they may not. It means that even if they existed, they wouldn't matter to the individual.

Imagine, for example, that after we'd gotten rid of gender, we discovered that more females than males participate in the International Mathematical Olympiad. Would we use this knowledge to discourage a math-loving male child from studying this subject, letting him know that he is wasting his time because only a few children with male genitals manage to excel in math in the long run? This might sound outrageous, but that's how we today commonly treat children with female genitals with respect to math.

When I talk about opening up such fields as math and computer science to more girls, and literature and the arts to more boys, I'm not calling on Red Riding Hood to save the hunter. I'm not telling children to do the opposite of their gender roles. Rather, I am suggesting we make a conscious effort to get rid of gender labels, so that children can develop into full human beings, instead of being forced into male or female pigeonholes.

With this in mind, I'd like to say a few words about single-sex education. It has long been practiced throughout the world, often for religious and traditional reasons. In the past decade or so, single-sex schooling has been generating growing interest due to the erroneous belief that boys and girls have different types of brains and therefore learn differently.[7]

Some proponents of single-sex education, however, raise genuine concerns—for example, that in a mixed school, a classroom runs the risk of being dominated by a few rowdy boys, making it difficult for girls to develop their skills. Yet when a group of dominant children takes over, *all* other children, boys and girls alike, have difficulty expressing themselves without becoming a target of ridicule or violence. Moreover, these dominant boys also find themselves in a bind, unable to deviate from behaviors dictated by their group. So creating a tolerant environment in which a boy "takeover" won't happen seems like a better solution for all children than creating a boy-free "safe zone" for girls.

Advocates of single-sex schooling sometimes also argue that it can help avoid stereotyping certain activities—chess or basketball for boys only, ballet and playing the violin for girls only, and so on. These advocates may be right, but this achievement is dwarfed by the unfortunate message conveyed by the mere existence of single-sex education: that one's sex category is *so* important, it even determines who attends which school. Even though it may be more difficult to get rid of gender labels in a mixed classroom, it's precisely the environment in which we can convey the crucial message—that the form of one's genitals has no social meaning.

Children do not belong to two types of learners, the boy type and the girl type, and separating them in the classroom by the form of their genitals is not the right way to go. We should do just the opposite: broaden teaching methods in schools so that they fit the needs of all children, encouraging them all to develop a variety of skills, both masculine and feminine—except that we will eventually stop using these terms, one hopes.

For example, children who have a hard time sitting for hours in the classroom would benefit from studying while engaging in physical activity—say, reciting the multiplication table while jumping with a rope. For children who much prefer sitting in class to being active, jumping with a rope could be excellent for developing their physical abilities. So having all children study the multiplication table with both methods would help them all master a crucial skill while giving them the opportunity to get better at what they find difficult, be it by nature or for other reasons—that is, sitting quietly or being physically active.

Building up a gender-free environment for children surely requires lots of effort—not only on the part of the teachers, but also on the part of the parents, who need to let go of gendered expectations from their children. In my kids' school, knitting is included in the curriculum from the first grade to the sixth, for all girls and boys. Every year, when the parents of first graders learn about this, a few fathers of boys raise objections, but, so far, they have refrained from inciting their children against the knitting. The children themselves take pride in mastering the new skill, having no clue it is viewed as a feminine activity (luckily, as knitting is so out of fashion in Israel, they cannot learn this from the media). I remember driving my soccer-loving son with two of his teammates back from soccer practice, and listening to them talking about their animal-knitting assignment. While they compared the number of stitches it took for different animals—one was knitting a cat, another a bear—I couldn't help picturing them one day, when they would become grandparents, knitting socks for their grandchildren.

Making our education system gender-free is a demanding task that may take time. But you can start making changes in people's lives now—starting, for example, with your own children.

20.

Ungendering
Our Children

A few years ago, my youngest son, then around five years old, came home from a birthday party all covered with pink pieces of paper, and announced that he was a pink ninja. He wandered happily around the house throwing pink paper ribbons at imaginary enemies, then came up to me and asked: "Do you know there are people who think boys don't like pink?" I said I knew that (and spared him the complete answer: that a hundred years ago, pink was for boys, and blue for girls). My son then stated: "These people are strange, because I'm a boy and I like pink."

That moment I realized that I had managed to help my children see the absurdity of the messages they were getting from society. My son didn't think *he* was strange, he knew it was the others. Indeed, how can one claim boys don't like pink, if there is even a single boy who does?

It turns out that children know this instinctively, even if society puts much effort into making them unknow it. Once, for example, I was riding in the car with my kids, all very young then, listening to a song on the radio in which the female singer sang about being "more of a woman than it seems." To me this sentence made perfect sense: bombarded as we are by messages about what is appropriate for men and women, aren't many of us worried about not being man or woman enough? But my kids asked me: What does the singer mean? How can a woman be more or less a woman?

I often hear from parents who try to raise their young children in a gender-free way—for example, by buying them *all* types of toys—how disappointed they are to discover that for all their efforts, the children had somehow managed to pick up gender stereotypes. Some parents take this to mean that the stereotypes must be "naturally" imprinted in us from birth. I respond to them that this may very well be true, but their children's coming up with the stereotypes out of thin air cannot serve as evidence. That's because children are immersed in gender stereotypes wherever they go. Indeed, studies have found that the extent to which children conform to gender norms is determined much more by their peers and the media than by their parents.

This is why I don't try to conceal gender stereotypes from my children. On the contrary, a good strategy is to acknowledge the existence of stereotypes and expose them as such—misguided beliefs that sometimes rest on a certain amount of evidence but often don't. For example, in Israel, the stereotype that girls are lousy at sports rests on the true observation that more boys than girls enroll in sports nationwide, but this observation surely does not mean that if you're a girl, you are no good at sports. On the other hand, the stereotype that girls are lousy at math is not backed by any kind of evidence: girls in Israel, on average, do at least as well at math as boys, in class and in international exams.

One way to expose a common belief as a stereotype rather than a truth about the world is to provide counterexamples. For instance, when my kids told me that girls don't know how to play soccer, I reminded them that it had been me who had taught them how to play. It worked. A few years later they had gone jogging with another family, but at some point their daughter stopped running, explaining that this was because she was a girl. To my delight, my kids declared that her explanation made no sense at all because they knew girls who ran.

We must let children know that the world still expects different things from boys and girls, and that some people are shortsighted. Otherwise, they might find out themselves the hard way. I recall my own painful experience when at around age ten I decided to join the swim team. I went there wearing the bottom part of a bathing suit, as I used to do on the beach. I didn't yet have breasts, and going like this to the beach was perfectly acceptable, but it wasn't on the swim team, a fact I realized by the stares of the girls and boys. I was mortified. For months afterward, I was sore at my mother for failing to educate me about other people's expectations, so that I could choose if and when to confront them.

To assist our children in detecting stereotypes, we can help them develop a critical eye: spotting messages aimed at channeling them onto a certain track and diverting them from what they want and need. When one side of a toy store is painted pink and the other blue, I point out to my kids how this color system, in effect, limits their choices. When we watch an action film together, I help them take note whether most of the active figures are men—as they usually are—and whether most of the ones who need to be rescued are women.

In addition to educating our children, we can try to affect their surroundings. When my oldest son was in second grade, I learned that he and his classmates were studying stories from the Bible, in which most prominent figures were male. They were also listening to stories about *tzadiks*—people considered righteous in Judaism—all of whom were male, and attending a reading of *The Wonderful Adventures of Nils*. I suggested to the teacher that she include in the readings stories focusing on female heroines. Once the bias was pointed out to her, she gladly did.

I find that people are often unaware of their own gender biases and are more open to change than one might think. When one of my children was in kindergarten, all the parents received an email inviting the children to a joint birthday party of four kids born that month. The email was signed by the four mothers. I used the "reply all" option to alert the entire mailing list to the amazing coincidence—that all the children celebrating a birthday that month didn't have a father. One of the fathers replied that, actually, his child did have a dad. But since that email, all invitations were signed by both parents or "the family of."

Our society in general places such a premium on motherhood that fathers are commonly kept on the margins of child care. They are often not included in a class's mailing list for parents; teachers commonly approach the mother, not the father, if they want to discuss the child, and so do other parents, for example, if they want to set a playdate. This approach creates a vicious circle of exclusion: it leaves the father uninformed, so that others feel it's a waste of time talking to him about the child, as he wouldn't know much anyway; and he himself finds it hard to be involved without making an extra effort to keep up with the child's life.

My kids' father is annoyed by this segregation at least as much as I am. He feels that it overlooks his important role as a father and resents being dependent on me for updates about *his* kids. Gaining recognition as a father in the workplace has also been a struggle for him. He had to turn down job offers on a number of occasions because his request to leave work early twice a week to take care of his kids was refused—a request that might have been granted more readily to a woman in his position.

Such a strict division of parenthood into gender roles ends up harming all parties involved—men, women, and children. Studies show that a father's involvement in his child's life plays a major role in the child's cognitive development and emotional health. A 2007 study published in *Applied Development Science* found, for example, that babies who spent more time alone with their fathers had fewer behavior problems later in childhood.[1] A 2015 study published in *The Scandinavian Journal of Economics* found that children whose fathers had taken a paternity leave did better in school years later.[2]

Other studies have found correlations between taking paternity leaves and fewer health problems for both mother and father, increased stability in the couple's relationship, and a more equal division of household chores.[3] In Iceland, for example, where fathers are entitled to a three-month paternity leave, research showed that three years later, care of the children was equally divided between mother and father in 63 percent of families in which the father had taken a paternity leave, compared with 41 percent of families in which only the mother had taken a leave.[4]

An effective way of making sure fathers take advantage of their right to a paternity leave is a policy known as the "daddy quota"—a paid leave reserved for fathers, in addition to a paid maternity leave. In Scandinavian countries, this policy has led to a huge increase in the percentage of fathers taking paternity leave, from often near zero to 80 to 90 percent. Even a short paternity leave has been shown to increase fathers' involvement with their child and to decrease the likelihood of the child having developmental problems later in life.[5]

21.

Gender Awareness

Each year I lead a students' group at my university to raise awareness of the gender system.[1] In one, a male participant identified himself as "genderqueer." He was handsome, bearded, and muscular and wore earrings, necklaces, and other jewelry. When one of the female students asked why he'd chosen to announce his gender identity in this manner, I directed the same question to her, asking why she'd chosen to advertise her own gender identity by wearing makeup and tight-fitting feminine outfits. She was surprised at first, then realized that, just as he was wearing a "genderqueer" sign, she was putting on a "woman" sign every morning.

The gender binary is so pervasive that, as this incident shows, it takes special effort to grasp to what extent it permeates our lives—from external trappings, which, in and of themselves, usually do not cause any harm, to perceptions and standards that do. Before we help others become aware of their gender biases and schemas, it's a good idea to take note of our own. Are gender norms constricting our own behavior?

There are societies in which you don't need to major in gender studies to know the answer. In parts of the world, girls and women risk their lives if they insist on getting an education. In many countries, including in the West, transgender and genderqueer persons commonly face violence merely for acting in a gender-nontypical way. But even in the safety of our home, when no one is watching, we sometimes avoid tasks we think of—often unconsciously—as "gender-inappropriate," such as assembling an Ikea bookcase or opening a cookbook to make soup.

It can be an eye-opening experience to have a look at polls revealing how our expectations differ for women and men. In 2017, the Pew Research Center asked 4,573 Americans to describe traits that they think society does and doesn't value for each gender.[2] The patterns that emerged from the answers matched the prevalent stereotypes. Traits related to strength and ambition were particularly valued for men; compassion and kindness, for women. For some of the words used to describe human qualities, Americans attached opposite values to men and women. For example, the word "powerful" was used positively in 67 percent of the cases when applied to describe men, but when applied to describe women, it was used negatively in 92 percent of the cases. "Emotional" was used almost exclusively to describe men—in a negative sense. "Promiscuous" was also used negatively, but applied primarily to women. These examples alone go to show what a great change is still required before we meet the goal I described earlier: that every behavior or trait be judged as to whether it is appropriate for humans.

Women, for example, pay a much heavier toll than men for being successful, especially if they succeed in areas traditionally reserved for men. Such women are viewed as tough and achievement-oriented, and consequently— because these qualities are deemed undesirable for women—less socially appealing. In a study published in the *Journal of Applied Psychology,* researchers from New York University and Columbia University asked participants to evaluate hypothetical assistant vice presidents for sales in an aircraft company, stereotypically a "man's job" because the products included engine assemblies and other aircraft equipment.[3] The same profiles of assistant VPs were labeled "Andrea" in some cases, and "James" in others. The researchers found that when the competence of the VPs was evaluated as ambiguous, study participants ranked Andrea and James as similarly likable. But when the VPs were designated as "stellar performers," Andrea came out as far less likable than James—she was, in effect, punished for her success. Remember this study next time you think a successful woman is a bitch.

Not only in the workplace but in a host of everyday situations, we commonly respond differently to similar behaviors, depending on whether they are displayed by women or men. Once, in one of my gender groups, a male participant walked in late and asked others what he'd missed. When a few of the women—who generally make up the majority in these groups—volunteered to fill him in, I stopped the class to draw the students' attention to what had just happened. I reminded them that a few minutes earlier, three women had been late; two had taken their seats quietly to avoid interfering with the class; one had asked what she'd missed, but no one had bothered to respond. Such commentary usually comes as a shock to both sides. Female students say they were sure they'd simply been nice to a fellow student, unaware how ungenerous they'd been to a fellow *female* student; male students are just as surprised to become aware of the "male privilege" exposed by this simple interaction—both in their own entitlement to disturb the entire group, and in taking for granted the positive response of others.

Discovering that you are being treated in a privileged way can be painful. Men in my groups, for example, are often distressed to discover that women give them lots of space and let them dominate conversations not because they are irresistibly charming, but because they are men.

One of the greatest privileges of being privileged is not knowing you have privilege. You may not even notice there's a step in a path you take every day until you see a person in a wheelchair being unable to get past that point. If you want to discover your own privileges, try to learn about the experiences of people who don't belong to your group. If you are a man, find out how women feel in various situations; if you are part of a Caucasian majority, learn about the experiences of minority groups; if you are able-bodied, talk to a person with a disability.

Once we become aware of gender biases and privileges, we can try to counteract them. One of my students was appalled to discover that at night, women may be afraid of passing him in the street because he is a man. His solution was to counteract one stereotype with another. Whenever he finds himself on a dark street and sees a woman in the distance, he crosses over to the opposite side and starts singing a pop song by Madonna. Because she is a gay icon, he hopes that the woman in the street might think he is gay and therefore consider him to be less of a threat.

The advantage of gender awareness is that it allows for choice. When women in my groups realize they acted in a caring and considerate way in certain circumstances because that's what's expected of women, this doesn't mean they'll stop being caring and considerate. It means they now have a choice whether to act in this manner, depending on whether the situation warrants it or not, rather than blindly following the dictates of gender.

Male participants in my groups also welcome the freedom that comes with abandoning gender roles. They say they want to have the choice whether to show physical or emotional strength, fix things, take the lead, or come to the rescue of others. One male student, for example, said he resented the fact that his girlfriend had expected him to change a flat tire in her car without even considering other solutions. He said he didn't mind the work itself and would have welcomed the same request from a male friend who needed help. What he minded was his girlfriend's unquestioning conviction that the flat tire was *his* responsibility.

Gender expectations also limit the choices in interactions with romantic partners. A female participant in one of my groups said she was amazed to discover how her interactions with her female friends, in which she often took the initiative, contrasted with the way she acted on a date with a man. When she'd gone on a first date a few days earlier, it was the man who'd come to pick her up, made suggestions as to what they should do, and decided where they'd go. The men in the group immediately stepped in to say how burdensome it was for them, being expected to always take the lead in dating.

A tool I find helpful in raising gender awareness is to imagine how a situation would play out if the participants' gender was reversed. On one occasion, when female and male students in the group were discussing sexual harassment, one of the men, who until then had not participated in the discussion, apologized for changing the subject and said he was more interested in scientific aspects of gender. Several of the women immediately invited him to bring up the issues that interested him. At this point, I drew the group's attention to how readily the women, despite being in the majority, had given up their power, agreeing to drop a subject that was relevant to many of them and one they seldom get a chance to discuss in a gender-mixed group. This exchange helped the students realize how being a man means feeling free to voice one's discomfort to the group; being a woman means making sure everyone present, especially men, feels comfortable. Imagine a situation in which men are enthusiastically discussing sports, then a woman asks them to change the subject, and they eagerly comply. Not that such a scenario is impossible, but it's highly unlikely.

Look for the footprints of gender in day-to-day situations. Is it mainly the men or the women who get up to help with the dishes after a family dinner? In the workplace, is it men or women who volunteer to perform unpaid duties, such as organizing an outing for all employees? During a business meeting, do ideas proposed by men and women receive the same attention?

Check yourself, too. When interviewing a prospective employee, discussing a mutual friend, or commenting on someone's driving, would your perception, words, or reactions have been the same had this person belonged to the other gender?

22.

Taking Action

Noticing the myriad ways in which the gender system operates can be frustrating, but it also means there are many things around us we can try to change. And it's worth trying.

Here's one inspiring example. "Orchestrating Impartiality," a study published in 2000 by Harvard University researchers, showed that when certain orchestras in the United States adopted "blind" auditions—conducted behind a screen that hid from the jury the identity of the applicant, including his or her gender—the number of female musicians hired by these orchestras increased sharply.[1] According to a summary of the study on Harvard Kennedy School's website, the percentage of female musicians in the five highest-ranked orchestras in the nation more than tripled in the years after this practice was introduced: from 6 to 21 percent.[2]

We cannot put up similar screens in most cases. But we can look for ways of eliminating gender as a relevant category.

One place to start might be removing the sex/gender question from the multitude of forms on which it appears for no good reason.[3] Why do border protection authorities need to know whether we are female or male when we fill out a customs declaration? Why do we have to check "Mr." or "Ms." when donating to a charity?

It's true that when it comes to documentation, we've come a long way compared to a couple of centuries ago. My coauthor reports that in working on a book about a nineteenth-century Russian biologist,[4] she had the hardest time finding archival information about his female ancestors, because women then weren't deemed important enough to figure in most official documents—thus the scientist's birth record from 1845 states his father's name and place of residence but makes no mention of his mother. But we have yet a long way to go. In Israel, for instance, it's not uncommon for applicants who are filling out official requests to be asked for the name of their father, though it's unclear why this information is needed and, even if it is, why father but not mother or, better still, parents (of whichever sex).

Our language also introduces the sex/gender category

much more often than is truly needed. Note that we seldom use physiological features other than sex to relate to people. Imagine how absurd it would be to say, "Here's my little brown-eyed kid. How was school today? Did you play with other brown-eyed kids or with blue-eyed ones?" Now imagine that you'd say, "Here's my little kid with male genitals. How was school today? Did you play with other kids with male genitals or with those with female genitals?" This surely sounds awful, but that's just what you say when you refer to your "little boy" or to other kids as "boys and girls," or when you talk about "men and women."

Whenever possible, you can try to avoid referring to human beings by their sex category, addressing them instead as "kids," "people," "folks," or "you guys," so that there's no connection between the form of their genitals and the message you want to convey. You can also try to use gender-neutral terms instead of job titles with male suffixes: "chair" instead of "chairman," or "police officer" instead of "policeman."[5]

To further free language from the current gender norms, pause to consider, why are men listed first in phrases referring to both sexes, as in "boys and girls" or "men and women"? If you think it goes by alphabetical order, what about "male and female"? And if you think the female-first phrases sound awkward, note that in this book, I've tried to alternate. I'd like to think that "women and men" and "female and male" by now sound just as harmonious to your ear as do their male-first alternatives.

When we can't cause gender to become invisible, understanding how it operates may be the most effective approach to fighting its biases. Consider a remarkable change that was brought about at Harvey Mudd College in Southern California in studies of computer science, a discipline with the lowest proportion of women receiving bachelor degrees of all the tech fields.[6] Harvey Mudd used to be no exception. When the faculty decided to remedy the situation, they first got at the root of this gender inequality; then they split the introductory computer science course into two sections, one for students with prior programming experience (most of them men) and one without (most of them women), to prevent students unfamiliar with programming from feeling intimidated. They also redesigned this course to include broader topics, such as benefiting society, rather than focusing on straight programming. "This worked wonders to create a supportive atmosphere," college president Maria Klawe wrote in *Newsweek*. The numbers followed suit. Klawe: "Within four years, we went from averaging around 10 percent women majors to averaging 40 percent. We have continued to average 40 percent since 2011."[7]

In another example, figuring out how gender distorts the organization of scientific conferences has enabled a colleague of mine to amend this process at least in some cases. She noticed that conference organizers, when putting together a list of potential speakers, were—almost certainly unintentionally—applying different criteria to women and men. When the name of a woman scientist was brought up, reservations were often raised: she wasn't a great speaker, or didn't work exactly in the field of the conference, or didn't have a sufficiently high status. Things were much easier with men. The organizing committee would start out by listing all the men who just *had* to be there: there was this big name who simply had to be invited, even though he wasn't exactly in the field, and this other guy, a terrible speaker who always gave the same talk but was a central figure in this area of studies, and this young man doing interesting research, it was time the field heard about his work.

With a bit of luck, the organizers would sometimes realize they didn't have enough women on the list and turn to "fill-in" women—women who are so well known they are the first to come to mind when one needs a token woman to be added to a men-only list of speakers. But these female scientists are so overwhelmed with invitations, some from outside their field, they often have to turn them down. (If you ever hear that a conference had no or few women speakers because they'd all refused to come, now you'll know why.) The result has been termed "a vicious circle of invisibility":[8] when you don't see women from your field, including promising young women, among speakers at a conference, you are unlikely to invite such women to the next conference.

When my colleague next organized a conference, she suggested reversing the order: first inviting all the relevant women speakers, then filling in the remaining slots with men, applying to them the same criteria typically applied to women—that they be great speakers *and* work exactly in the defined field. The question the organizers asked themselves was: "If this (male) researcher were a woman, would we have invited him?" It turned out to be a superb conference for all the women and men who took part.

To help break the vicious circle of women's invisibility at conferences, several scientists, women and men, have created a website, BiasWatchNeuro, that lists the percentages of women and men among invited speakers at conferences in various areas of neuroscience and compares these percentages to the base rates of women and men in each field. For example, if women make up 40 percent of scientists in a particular field but only 20 percent of the invited speakers at a conference, this means that they are underrepresented at this conference.

The website has made a difference. Whether it's because it's made conference organizers aware of gender bias or because they don't want to be shamed for not including enough women, they've been trying harder to include more women speakers. And guess what? They have been succeeding. Among the major success stories highlighted at BiasWatchNeuro is the Bernstein

Conference, the largest annual computational neuroscience conference in Europe. From 2009 to 2015, invited speakers and program chairs at these meetings were predominantly men. But in 2016, 42 percent of invited speakers were women—above the base rate. In fact, by the end of 2018, women were represented above their base rate in the majority of evaluated conferences.[9] BiasWatchNeuro has also inspired at least two similar websites: BiasWatchArchaeo in archeology and BiasWatchDerm in dermatology.

Fighting gender bias is, of course, easier when the initiative comes from above. Few measures can probably bring about societal change more effectively than legislation. One oft-cited example is the impact of Title IX—a US law passed in 1972 banning discrimination in federally funded educational activities on the basis of sex—on opportunities for girls and women in sports. In 1971, just before Title IX was enacted, there was only 1 girl for every 12 boys participating in high-school athletics; by 2013, this ratio stood at 1 girl for every 1.39 boys. In colleges that belong to the National Collegiate Athletic Association, there were only 2.5 women's teams per school prior to the enactment of Title IX; by 2014, this number more than tripled, reaching an average of 8.83 women's teams per school.[10]

In Israel, a government initiative has led to a dramatic increase in the number of women on boards of public and government companies. Before 1993, women made up

173

only 7 percent of such board members, but after a correction about appropriate gender representation was made that year in the government company law, their number rose to 29 percent in 1997 and 37.8 percent in 2000. Then, in 2007, the government passed a decision that "appropriate representation" meant 50 percent. By 2013, women made up 45.5 percent of board members.[11]

In September 2018, California became the first state in the United States to require that at least one woman be included on the board of directors of all publicly traded California companies. One hopes it's only a first step toward a more balanced gender ratio than the current one of over five men to every woman.[12]

Although a small change in policy or legislation can lead to a large societal shift, making such changes often requires a massive collective effort. But it takes far less effort to simply make a difference. In you. In your loved ones. In the people around you.

It's true that fighting the gender system can be fraught. Commenting on gendered expressions among friends or at your workplace can earn you the label "annoying feminist," which often goes with "man hater" if you are a woman. Protesting against sexist jokes may not only tag you as humorless but expose you to verbal or physical aggression. On the other hand, many people are happy to discover their own and others' implicit biases, especially if you point these out to them in private and in a non-judgmental manner (nasty comments on their Facebook wall are less likely to be welcomed). And if in the course of these revelations you become aware of the power or privileges granted to you by the gender system, why not use this power and these privileges to try to eliminate this system from our lives?

23.

Vision

I'd like to end with a few words about my vision.

I see a future in which there are no men and women—only humans with female, male, or intersex genitals. In this vision, sex is just a term to describe one of our physical characteristics—like height, weight, age, or the color of our eyes—but it is not used to divide humans into groups and treat them differently. It says nothing about whether they are supposed to like football or poetry.

If this future is difficult to imagine, think of handedness. Some people are right-handed; some are left-handed; some are right-handed in certain tasks, left-handed in others.

Not so long ago, left-handed people were thought of as less capable physically and mentally than right-handed people. Our language still carries traces of that time in such phrases as "She has two left hands." In the 1948 hit musical *Easter Parade*, the Judy Garland character says she was born left-handed, so they'd tied up her left hand, and the doctor told her mother she might grow up to be a dangerous criminal. At the time the musical was made, this story was told only partly in jest. When parents and teachers noted that a child was using her or his left hand, they would tie up that hand, to make the child use the right hand. Numerous studies were conducted to look for brain deficits responsible for the purported inferiority of left-handed people. One theory held that left-handedness results from a disrupted development of the brain's left hemisphere, which controls the right side of the body, and this supposedly causes the left hand to become dominant instead of the *right* one. (Note the unintended pun in the previous sentence, again suggesting the left hand is the wrong one.)

Today, although there are still right- and left-handed individuals, handedness carries no meaning beyond the description of a physiological characteristic. This characteristic is important in some situations. For example, if I want to buy scissors, I need to know whether to buy the ones for right- or left-handed people, or if I'm going to play tennis, it may be wise to know if my opponent is right- or left-handed. But I don't care one bit if my physician or my child's teacher is right- or left-handed. I never ask my kids whether they played with right- or left-handed children in the playground. It makes no difference to you that this text has been written by two right-handed persons.

That is the way we should treat sex, too. Why should we care whether our physician, our children's teacher, or their playmate has male or female genitals? Why should it make a difference to you that this text has been written by authors with female genitals?

Sex category can surely be important in certain situations, even crucial when we intend to have a child. But in a gender-free world, the pressure to create a heteronormative family—a male, a female, and their biological children—will be lifted. If we stop being obsessed with the form of one's genitals, why should we care about it in relationships? A heteronormative family is a human invention, not a relationship-cum-genitals package deal that emerged in the course of evolution. When it comes to raising children, the heteronormative family is just one of several ways to go about it.[1]

In a world without gender, the sex categories themselves will lose their centrality. Devoid of social significance, the form of a person's genitals will be unimportant. This will be a great relief to people with atypical genitals, and certainly to parents of babies with such genitals, as the pressure to subject their children to surgical procedures would be alleviated.

Instead of squeezing the enormous human variety on all possible traits into two boxes, we will celebrate this variability. Today only a minority of humans identify themselves using descriptors other than the traditional terms "woman" and "man," but in a world without a binary gender system, many more would resort to defining themselves through multiple gender categories. I'm not sure we'd continue calling them *gender* categories, though.

One objection that's been raised at a lecture of mine is that even without gender, women and men would still behave differently because they differ biologically. I see no problem with that. On the contrary, if we believe that biology would drive the behavior of females and males apart, there's surely no reason to introduce all those gender conventions to achieve the same end.

A concern I often hear about a world without gender is that it will be boring because everybody will be the same. I believe the opposite is true. We are not the same and won't be after binary gender disappears or gives way to multiple identities. In fact, the more, the merrier.

Just ponder all the variations on the "boy meets girl" theme in a world with more categories. Liberated from gender binaries, scriptwriters will draw on dozens of different gender identities—think "pangender meets gender-undecided"—highlighting each one's characteristic quirks. The number of potential combinations will go through the roof—a possibility I find anything but boring.

Yet another concern I occasionally hear is that with everyone dressing and behaving however they want, it would be impossible to tell a female from a male. How would they know whom to date, people ask. My suggestion is to date only people to whom you are attracted. For example, if you think you are attracted only to females, it shouldn't matter how they are dressed. And if you find yourself attracted to someone and discover they have male genitals, then I guess your original assumption was wrong.

Gender is a system that assigns meaning to sex, but our gender is not a reflection of our sex. Gender is one of the prisons within which we live. It divides the world into things for males and things for females. And if we want things that are not on "our" side, we are punished by society.

In the world I envision, there is no gender.

There is only sex.

Humans with female, male, or intersex genitals are free to choose from all that this world has to offer. Some will choose only dolls, others will choose only balls, many will choose both. Whatever you love and do, if it's appropriate for humans, it is appropriate for you.

Acknowledgments

DAPHNA JOEL

This book would not have been born without Luba Vikhanski. I'd long wanted to write a book that would make my brain mosaic concept known outside the scientific community, but wasn't sure how to go about it. In Luba, I found a creative and caring writing partner who gave my ideas new shape and form to make them (I hope) accessible to a wide audience.

I'd come up with the brain mosaic idea while reading about sex and the brain, but developing it in full took a great deal of research, for which I relied on the cooperation of many people. I'm grateful to all those who accompanied me in these exciting explorations, and would like to acknowledge several people in particular.

I owe a special debt of gratitude to Ina Weiner, who introduced me to scientific research while serving as my PhD adviser, taught me to write scientific papers, and became a cherished colleague and friend. She was my sounding board throughout the work on this book, offering helpful advice and making excellent comments on the manuscript in its various versions.

Members of the NeuroGenderings Network were incredibly welcoming in inviting me to join their ranks and opening their minds and hearts to my ideas on sex and the brain. I greatly appreciate their critical thinking, humor, and readiness to address any question I could possibly raise.

Among members of the Network, special thanks go to Anne Fausto-Sterling, for all that I've learned from her books (see Suggested Reading), for agreeing to write with me a joint paper about my work, and for her willingness to provide advice whenever I needed it. I'd also like to extend special thanks to Cordelia Fine, for the joy that I find in collaborating with her on scientific and popular articles, and for her sensitive critiques of my own writings, including this book. I invariably find great pleasure in exchanging ideas and collaborating with Gina Rippon, Rebecca Jordan-Young, and Anelis Kaiser, and I'm grateful to Cordelia, Gina, Rebecca, and Anelis for our friendship.

I want to thank Marcia Stefanick for recognizing the potential impact of the mosaic hypothesis early on, for opening the door for me into the community of sex differences researchers, and for having been a supportive colleague and friend ever since. Of the many sex differences researchers who were willing to integrate my ideas with theirs, I want to thank, in particular, Margaret (Peg) McCarthy, for our discussions of sex and the brain, which culminated in a joint paper, and Dick Swaab, for taking up with me a joint reanalysis of his data on sex differences in the human hypothalamus.

My research on the brain and gender mosaic and on gender identity would have been impossible without numerous collaborators in Israel and elsewhere, who shared with me their knowledge, experience, and data. My thanks go to Ricardo Tarrasch, Effi Ziv, Maya Mukamel, Yaniv Assaf, Sabine Oligschläger, Jared Pool, Sebastian Urchs, Daniel Margulies, Franziskus Liem, Jürgen Hänggi, Lutz Jäncke, Roee Admon, and Talma Hendler. I am especially grateful to Isaac Meilijson, Amir Averbuch, and Moshe Salhov for helping me find the right mathematical tools for testing my brain mosaic hypothesis.

I'm infinitely grateful to my students—in particular, to Zohar Berman, Ariel Persico, Roi Jacobson, Nadav Wexler, and Guy Shalev—for sharing my curiosity about sex, gender, and the brain and for helping me develop new ways of looking at the human brain and at humans in general.

My family and friends made my cause their own when I switched research fields to start working on sex, gender, and the brain. They incorporated my ideas into their lives, sharing with me how these ideas affected them and others, and encouraged me to write this book. I'm especially grateful to my parents, who for many years have been learning about the brain, and then gender, in order to understand my research, who never miss an interview with me or an article about my work, and who are always there for me. My sister has for years been a source of strength and loving support. I'm grateful to my three kids for sharing my excitement about every new research result, for all that I've learned and continue learning from them, and for their love.

LUBA VIKHANSKI

When Daphna Joel first contacted me with the idea for this book, I thought I was well prepared. About ten years earlier, another scientist had asked me to write a book with him, and amazingly enough, it had been on the same topic: sex and the brain. That project never took off, but while considering it I'd spent several months reading up on the subject. Still, it took quite a learning process for me to become fully converted to the mosaic view of sex, gender, and the brain. I'm grateful to Daphna for this fascinating intellectual adventure.

Going further back in time, I'd like to thank several people for playing critical roles in helping me become a science writer.

When I was growing up in Moscow, then behind the Iron Curtain, all foreign movies were dubbed into Russian, access to books in English was limited, and having contacts with foreigners was dangerous. Despite these linguistically sterile conditions, two wonderful teachers, Inna Aronovna Cohn and the late Anna Pavlovna Maslova, managed to help me master English, the language in which I write today.

Years later, after I'd moved to Israel, author and editor Mikhail Heifetz gave me my first writing assignment: a review for a Jerusalem-based magazine.

William E. Burrows, a.k.a. "Mom," founded New York University's Science, Health and Environmental Reporting Program, which enabled me to get started in science writing.

Jane Nevins, then editor in chief of the Dana Press, gave me confidence that I could write books.

I was fortunate to benefit from mentoring by the late Rinna Samuel, who for a number of years critiqued my work at my request, sharing with me her remarkable editorial wisdom.

DAPHNA AND LUBA

We owe a great debt to Eva Illouz, who introduced us to each other, and are grateful to her for providing insightful feedback on the manuscript in its early stages.

Several other friends and colleagues generously gave of their time, reading various versions of the entire manuscript and making no end of valuable comments. In particular, we'd like to thank Baat Enosh, Eda Goldstein, Navah Haber-Schaim, and Riza Jungreis.

In Deborah Harris, we found the agent of our dreams. She kept an open mind when at first we wanted to write an ultrashort book, stood behind the project when we expanded the manuscript to its current length, and followed it through with impressive expertise in every possible aspect of publishing.

At Little, Brown, we thank Tracy Behar for her early enthusiasm and Marisa Vigilante for her sharp editorial comments. Janet Byrne did a fantastic copyediting job on the manuscript. The entire team at Little, Brown took us through the book production process in an efficient, professional, and enjoyable manner.

Notes

PART I. SEX AND THE BRAIN

Chapter 1. My Awakening

1. Marwha, D., M. Halari, and L. Eliot. 2017. Meta-analysis reveals a lack of sexual dimorphism in human amygdala volume. *Neuroimage* 147:282–94.

Chapter 2. A History of Twisted Facts

1. Quoted in Schiebinger, L. 1989. *The Mind Has No Sex? Women in the Origins of Modern Science.* Cambridge, MA: Harvard University Press, 217.
2. Ibid.
3. Ibid., 189.
4. Ibid., 215.
5. Shields, S. A. 1975. Functionalism, Darwinism, and the psychology of women: A study in social myth. *American Psychologist* 30:739–54.
6. Quoted in Blum, D. 1997. *Sex on the Brain.* New York: Penguin Books, 38.
7. Romanes, G. J. 1887. Mental differences between men and women. *Nineteenth Century* 21:654–72. Quoted in Shields, Functionalism, Darwinism, and the psychology of women.
8. Mazón, P. M. 2003. *Gender and the Modern Research University:*

The Admission of Women to German Higher Education, 1865–1914. Palo Alto, CA: Stanford University Press, 89.

9. Shields, Functionalism, Darwinism, and the psychology of women.

10. Ibid.

11. Patrick, G. T. W. 1895. The psychology of women. *Popular Science Monthly* 47:212. Quoted in Shields, Functionalism, Darwinism, and the psychology of women.

12. Schiebinger, *The Mind Has No Sex?*, 2.

Chapter 3. As the Differences Pile Up

1. The 56-minute discussion is available at www.youtube.com/watch?v=64fPSH5qqS4. Accessed January 17, 2019.

2. Ruigrok, A. N. V., et al. 2014. A meta-analysis of sex differences in human brain structure. *Neuroscience & Biobehavioral Reviews* 39:34–50.

3. Ritchie, S. J., et al. 2018. Sex differences in the adult human brain: Evidence from 5,216 UK Biobank participants. *Cerebral Cortex* 28(8):2959–75.

4. Ibid.

5. Shaywitz, B. A., et al. 1995. Sex differences in the functional organization of the brain for language. *Nature* 373(6515):607–9.

6. Sommer, I. E. 2008. Sex differences in handedness, asymmetry of the planum temporale and functional language lateralization. *Brain Research* 1206:76–88.

Chapter 4. Nature vs. Nurture

1. Gilmore, J. H., R. C. Knickmeyer, and W. Gao. 2018. Imaging structural and functional brain development in early childhood. *Nature Reviews–Neuroscience* 19(3):123–37.

2. Maguire, E. A., K. Woollett, and H. J. Spiers. 2006. London taxi drivers and bus drivers: A structural MRI and neuropsychological analysis. *Hippocampus* 16:1091–1101.

3. Karni, A., et al. 1998. The acquisition of skilled motor

performance: Fast and slow experience-driven changes in primary motor cortex. *Proceedings of the National Academy of Sciences, USA* 95(3):861–68.

4. Turkeltaub, P. E., et al. 2003. Development of neural mechanisms for reading. *Nature Neuroscience* 6(7):767.

5. Goffman, E. 1979 [1976]. *Gender Advertisements*. London: The Macmillan Press Ltd., 46.

6. Ibid., 8.

7. Richardson, S. S. 2013. *Sex Itself: The Search for Male and Female in the Human Genome*. Chicago: University of Chicago Press.

8. Hyde, J. S., et al. 2018. The future of sex and gender in psychology: Five challenges to the gender binary. *American Psychologist* 74:171–93; van Anders, S. M. 2013. Beyond masculinity: Testosterone, gender/sex, and human social behavior in a comparative context. *Frontiers in Neuroendocrinology* 34:198–210.

9. Van Anders, Beyond masculinity.

10. Abraham, E., et al. 2014. Father's brain is sensitive to childcare experiences. *Proceedings of the National Academy of Sciences, USA* 111(27):9792–97.

PART II. THE HUMAN MOSAIC

Chapter 5. Brains in Flux

1. Shors, T. J., C. Chua, and J. Falduto. 2001. Sex differences and opposite effects of stress on dendritic spine density in the male versus female hippocampus. *Journal of Neuroscience* 21(16):6292–97.

2. Joel, D., and M. M. McCarthy. 2017. Incorporating sex as a biological variable in neuropsychiatric research: Where are we now and where should we be? *Neuropsychopharmacology* 42(2):379–85. See figure 2 for a summary of direct and indirect effects of sex.

3. Shors, T. J. 2002. Opposite effects of stressful experience on memory formation in males versus females. *Dialogues in Clinical Neuroscience* 4(2):139–47.

4. Juraska, J. M., et al. 1985. Sex differences in the dendritic branching of dentate granule cells following differential experience. *Brain Research* 333(1):73–80.

5. Joel, D. 2012. Genetic-gonadal-genitals sex (3G-sex) and the misconception of brain and gender, or, why 3G-males and 3G-females have intersex brain and intersex gender. *Biology of Sex Differences* 3:27.

6. Reich, C. G., M. E. Taylor, and M. M. McCarthy. 2009. Differential effects of chronic unpredictable stress on hippocampal CB1 receptors in male and female rats. *Behavioural Brain Research* 203(2):264–69.

7. Garcia-Falgueras, A., et al. 2011. Galanin neurons in the intermediate nucleus (InM) of the human hypothalamus in relation to sex, age, and gender identity. *Journal of Comparative Neurology* 519(15):3061–84.

8. Chung, W. C. J., G. J. De Vries, and D. F. Swaab. 2002. Sexual differentiation of the bed nucleus of the stria terminalis in humans may extend into adulthood. *Journal of Neuroscience* 22(3):1027–33.

Chapter 6. Not by Sex Alone

1. Joel, D. 2011. Male or female? Brains are intersex. *Frontiers in Integrative Neuroscience* 5:57; Joel, D. 2012. Genetic-gonadal-genitals sex (3G-sex) and the misconception of brain and gender, or, why 3G-males and 3G-females have intersex brain and intersex gender. *Biology of Sex Differences* 3:27.

2. Juraska, J. M. 1991. Sex differences in "cognitive" regions of the rat brain. *Psychoneuroendocrinology* 16(13):105–19.

3. Yankelevitch-Yahav, R. The effects of postnatal administration of the selective serotonin reuptake inhibitor fluoxetine on female and male rats. PhD diss. under the supervision of D. Joel, Tel Aviv University, 2018.

4. Koss, W. A., et al. 2015. Gonadectomy before puberty increases the number of neurons and glia in the medial prefrontal

cortex of female, but not male, rats. *Developmental Psychobiology* 57(3):305–12.

5. McCarthy, M. M., and A. P. Arnold. 2011. Reframing sexual differentiation of the brain. *Nature Neuroscience* 14(6):677–83; McCarthy, M. M., et al. 2015. Surprising origins of sex differences in the brain. *Hormones and Behavior* 76:3–10.

6. Cahill, L. April 1, 2014. "Equal ≠ The Same: Sex Differences in the Human Brain." *Cerebrum.* http://dana.org/Cerebrum/2014/ Equal_%E2%89%A0_The_Same__Sex_Differences_in_the_Human_Brain/. Accessed January 17, 2019.

7. Fine, C., et al. December 15, 2014. "Why Males ≠ Corvettes, Females ≠ Volvos, and Scientific Criticism ≠ Ideology." Reaction to "Equal ≠ The Same: Sex Differences in the Human Brain." *Cerebrum.* http://dana.org/Cerebrum/Default.aspx?id=115816. Accessed January 17, 2019.

Chapter 7. Mosaic of the Human Brain

1. Joel, D., et al. 2015. Sex beyond the genitalia: The human brain mosaic. *Proceedings of the National Academy of Sciences, USA* 112(50):15468–73.

2. Ingalhalikar, M., et al. 2014. Sex differences in the structural connectome of the human brain. *Proceedings of the National Academy of Sciences, USA* 111(2):823–28.

3. Sample, I. December 2, 2013. "Male and Female Brains Wired Differently, Scans Reveal." *Guardian.* www.theguardian.com/ science/2013/dec/02/men-women-brains-wired-differently. Accessed January 17, 2019.

4. Joel, Sex beyond the genitalia, 15468.

5. For responses of other scientists to my research, see Denworth, L. 2017 (September). "Is there a 'female' brain?" *Scientific American* 317(3):43. This article appeared in a special issue of *Scientific American*, "The New Science of Sex and Gender."

Chapter 8. Now You See It, Now You Don't

1. Jordan, K., et al. 2002. Women and men exhibit different cortical activation patterns during mental rotation tasks. *Neuropsychologia* 40(13):2397–2408.
2. Weiss, E., et al. 2003. Sex differences in brain activation pattern during a visuospatial cognitive task: A functional magnetic resonance imaging study in healthy volunteers. *Neuroscience Letters* 344(3):169–72.
3. Hugdahl, K., T. Thomsen, and L. Ersland. 2006. Sex differences in visuo-spatial processing: An fMRI study of mental rotation. *Neuropsychologia* 44(9):1575–83.
4. Halari, R., et al. 2006. Comparable fMRI activity with differential behavioural performance on mental rotation and overt verbal fluency tasks in healthy men and women. *Experimental Brain Research* 169(1):1–14.
5. Carrillo, B., et al. 2010. Cortical activation during mental rotation in male-to-female and female-to-male transsexuals under hormonal treatment. *Psychoneuroendocrinology* 35(8):1213–22.
6. Jordan-Young, R. M. 2011. *Brain Storm: The Flaws in the Science of Sex Differences.* Cambridge, MA: Harvard University Press.
7. David, S. P., et al. 2018. Potential reporting bias in neuroimaging studies of sex differences. *Scientific Reports* 8, article number 6082.

Chapter 9. In Anticipation of a Blind Date

1. Eagly, A. H., and W. Wood. 2013. The nature–nurture debates: 25 years of challenges in understanding the psychology of gender. *Perspectives on Psychological Science* 8(3):340–57.

Chapter 10. Brain "Types," Common and Rare

1. Baron-Cohen, S. 2002. The extreme male brain theory of autism. *Trends in Cognitive Sciences* 6(6):248–54; Greenberg, D. M., et al. 2018. Testing the Empathizing–Systemizing theory of sex

differences and the Extreme Male Brain theory of autism in half a million people. *Proceedings of the National Academy of Sciences, USA* 115(48):12152–57.

2. Joel, D., et al. 2018. Analysis of human brain structure reveals that the brain "types" typical of males are also typical of females, and vice versa. *Frontiers in Human Neuroscience* 12:399.

3. GenderSci blog. October 29, 2018. A Q&A with Daphna Joel on her new article, "Analysis of Human Brain Structure Reveals That the Brain 'Types' Typical of Males Are Also Typical of Females, and Vice Versa." https://projects.iq.harvard.edu/gendersci/joelqa. Accessed January 17, 2019.

4. Koppel, M., et al. 2002. Automatically categorizing written texts by author gender. *Literary and Linguistic Computing* 17(4):401–12.

Chapter 11. Men and Women Under Stress

1. www.stressinamerica.org. Accessed January 24, 2019.

2. Admon R., et al. 2013. Stress-induced reduction in hippocampal volume and connectivity with the ventromedial prefrontal cortex are related to maladaptive responses to stressful military service. *Human Brain Mapping* 34:2808–16.

3. Joel, D. In press. Beyond sex differences and a male-female continuum: Mosaic brains in a multidimensional space. In Lanzenberger, R., G. S. Kranz, and I. Savic, eds. *Handbook of Clinical Neurology,* 3rd Series, Sex Differences in Neurology and Psychiatry. Amsterdam: Elsevier.

4. "Gender and Stress." www.apa.org/news/press/releases/stress/2010/gender-stress. Accessed January 22, 2019.

5. "Stress in America™ Press Room." www.apa.org/news/press/releases/stress/index. Accessed March 20, 2019.

Chapter 12. The Human Health Mosaic

1. Clayton, J. A., and F. S. Collins. 2014. NIH to balance sex in cell and animal studies. *Nature* 509(7500):282–83.

2. Prendergast, B. J., K. G. Onishi, and I. Zucker. 2014. Female mice

liberated for inclusion in neuroscience and biomedical research. *Neuroscience & Biobehavioral Reviews* 40:1–5; Becker, J. B., B. J. Prendergast, and J. W. Liang. 2016. Female rats are not more variable than male rats: A meta-analysis of neuroscience studies. *Biology of Sex Differences* 7:34.

3. Clayton and Collins, NIH to balance sex.

4. For example, "Every Cell Has a Sex" is the title of Part II in Weizmann, T. A., and M. L. Pardue, eds. 2001. *Exploring the Biological Contributions to Health: Does Sex Matter?* Washington, DC: National Academy Press.

5. Joel, D., and R. Yankelevitch-Yahav. 2014. Reconceptualizing sex, brain and psychopathology: Interaction, interaction, interaction. *British Journal of Pharmacology* 171(20):4620–35; Hyde, J. S., et al. 2018. The future of sex and gender in psychology: Five challenges to the gender binary. *American Psychologist* 74:171–93.

6. Gillies, G. E., and S. McArthur. 2010. Estrogen actions in the brain and the basis for differential action in men and women: A case for sex-specific medicines. *Pharmacological Reviews* 62:155–98.

7. Hyde, The future of sex and gender.

8. Van Anders, S. M. 2013. Beyond masculinity: Testosterone, gender/sex, and human social behavior in a comparative context. *Frontiers in Neuroendocrinology* 34:198–210.

9. Coquelin, A., and C. Desjardins. 1982. Luteinizing hormone and testosterone secretion in young and old male mice. *American Journal of Physiology* 243(3):E257–63.

10. Fausto-Sterling, A., and D. Joel. December 6, 2017. The Science of Difference: Let's Do It Right! *Huffington Post*. www.huffingtonpost.com/dr-anne-fausto-sterling/the-science-of-difference-lets-do-it-right_b_5372859.html. Accessed January 26, 2019.

11. Joel, D., and A. Fausto-Sterling. 2016. Beyond sex differences: New approaches for thinking about variation in brain structure and function. *Philosophical Transactions of the Royal Society B* 371:20150451; Joel, D., and M. M. McCarthy. 2017.

Incorporating sex as a biological variable in neuropsychiatric research: Where are we now and where should we be? *Neuropsychopharmacology* 42(2):379–85.

12. Stefanick, M. L. 2017. "Not Just for Men." *Scientific American* 317(3):52.

13. Richardson, S. S., et al. Opinion: Focus on preclinical sex differences will not address women's and men's health disparities. *Proceedings of the National Academy of Sciences, USA* 112(44):13419–20.

14. Linde, C., et al. 2018. The interaction of sex, height, and QRS duration on the effects of cardiac resynchronization therapy on morbidity and mortality: An individual-patient data meta-analysis. *European Journal of Heart Failure* 20:780–91.

15. Canto, J. G., and C. Kiefe. 2014. Age-specific analyses of breast cancer versus heart disease mortality in women. *The American Journal of Cardiology* 113(2):410–11.

16. Cheng, E. R., et al. 2018. Prevalence of depression among fathers at the pediatric well-child care visit. *JAMA Pediatrics* 172(9):882–83.

Chapter 13. Mosaic of the Mind

1. Joel, D., et al. 2015. Sex beyond the genitalia: The human brain mosaic. *Proceedings of the National Academy of Sciences, USA* 112(50):15468–73.

2. Hyde, J. S. 2005. The gender similarities hypothesis. *American Psychologist* 60(6):581–92; Hyde, J. S. 2014. Gender similarities and differences. *Annual Review of Psychology* 65:373–98; Zell, E., Z. Krizan, and S. R. Teeter. 2015. Evaluating gender similarities and differences using metasynthesis. *American Psychologist* 70(1):10–20.

3. Carothers, B. J., and H. T. Reis. 2013. Men and women are from Earth: Examining the latent structure of gender. *Journal of Personality and Social Psychology* 104(2):385–407.

PART III. WHAT'S WRONG WITH GENDER

Chapter 14. From the Binary to a Mosaic

1. Terman, L. M., and C. C. Miles. 1936. *Sex and Personality: Studies in Masculinity and Femininity.* New York: McGraw-Hill, 1.
2. Ibid., 5.
3. Bem, S. L. 1974. The measurement of psychological androgyny. *Journal of Consulting and Clinical Psychology* 42(2):155–62.
4. Spence, J. T., R. Helmreich, and J. Stapp. 1975. Ratings of self and peers on sex role attributes and their relation to self-esteem and conceptions of masculinity and femininity. *Journal of Personality and Social Psychology* 32(1):29–39.
5. Spence, J. T. 1993. Gender-related traits and gender ideology: Evidence for a multifactorial theory. *Journal of Personality and Social Psychology* 64(4):624–35.

Chapter 15. Gender Illusions

1. Henley, N. M., and M. LaFrance. 1984. Gender as culture: Difference and dominance in nonverbal behavior. In Wolfgang, A., ed. *Nonverbal Behavior: Perspectives, Applications, Intercultural Insights,* 351–71. Ashland, OH: Hogrefe & Huber Publishers.
2. Hecht, M. A., and M. LaFrance. 1998. License or obligation to smile: The effect of power and sex on amount and type of smiling. *Personality and Social Psychology Bulletin* 24(12):1332–42.
3. Berg, J. H., W. G. Stephan, and M. Dodson. 1981. Attributional modesty in women. *Psychology of Women Quarterly* 5(5)Supplement:711–27; Gould, R. J., and C. G. Slone. 1982. The "feminine modesty" effect: A self-presentational interpretation of sex differences in causal attribution. *Personality and Social Psychology Bulletin* 8(3):477–85.
4. Eagly, A. H., W. Wood, and L. Fishbaugh. 1981. Sex differences in conformity: Surveillance by the group as a determinant of

male nonconformity. *Journal of Personality and Social Psychology* 40(2):384–94.

5. McGlone, M. S., and J. Aronson. 2006. Stereotype threat, identity salience, and spatial reasoning. *Journal of Applied Developmental Psychology* 27(5):486–93.

6. A highly effective exploration of the colors associated with baby girls and boys is provided by artist JeongMee Yoon in The Pink & Blue Project, www.jeongmeeyoon.com/aw_pinkblue_pink001.htm. Accessed January 26, 2019.

7. Shaked herself banked on her own good looks in a 2019 election video that mocked accusations of extremism against her party, Israel's New Right. The video showcased Shaked as a fashion model promoting Fascism perfume, www.youtube.com/watch?v=kLlnZGj83vM. Accessed March 24, 2019.

Chapter 16. Binary Brainwashing

1. Joel, D., and I. Weiner. 1994. The organization of the basal ganglia-thalamocortical circuits: Open-interconnected rather than closed segregated. *Neuroscience* 63:363–79.

2. Moss-Racusin, C. A., et al. 2012. Science faculty's subtle gender biases favor male students. *Proceedings of the National Academy of Sciences, USA* 109(41):16474–79.

3. Seavey, C. A., P. A. Katz, and Z. S. Rosenberg. 1975. Baby X: The effect of gender labels on adult responses to infants. *Sex Roles* 1(2):103–9.

4. Lavy, V., and E. Sand. 2018. On the origins of gender gaps in human capital: Short- and long-term consequences of teachers' biases. *Journal of Public Economics* 167:263–79.

5. Bernstein, H. August 1, 1909. "Metchnikoff—'the Apostle of Optimism'—on the Science of Living." *New York Times,* SM4.

6. Bian, L., S.-J. Leslie, and A. Cimpian. 2017. Gender stereotypes about intellectual ability emerge early and influence children's interests. *Science* 355(6323):389–91.

7. Cimpian, A., and S.-J. Leslie. September 2017. "The Brilliance Trap." *Scientific American* 317(3):61–65.

8. Leslie, S.-J., et al. 2015. Expectations of brilliance underlie gender distributions across academic disciplines. *Science* 347(6219): 262–65.

9. This issue is addressed in the *APA Guidelines for Psychological Practice with Boys and Men,* issued by the American Psychological Association in 2018. www.apa.org/monitor/2019/01/ce-corner.aspx?wpisrc=nl_lily&wpmm=1. Accessed January 22, 2019.

10. Eliot, L. 2009. *Pink Brain, Blue Brain: How Small Differences Grow into Troublesome Gaps—and What We Can Do About It.* Boston: Houghton Mifflin Harcourt.

11. Wong, Y. J., et al. 2017. Meta-analyses of the relationship between conformity to masculine norms and mental health-related outcomes. *Journal of Counseling Psychology* 64(1): 80–93.

PART IV. TOWARD A WORLD WITHOUT GENDER

Chapter 17. How to Deal with Gender Myths

1. Conger, K. August 5, 2017. "Exclusive: Here's the Full 10-Page Anti-Diversity Screed Circulating Internally at Google." *Gizmodo.* https://gizmodo.com/exclusive-heres-the-full-10-page-anti-diversity-screed-1797564320. Accessed January 28, 2019.

2. Su, R., J. Rounds, and P. I. Armstrong. 2009. Men and things, women and people: A meta-analysis of sex differences in interests. *Psychological Bulletin* 135(6):859–84.

3. Hyde, J. S. 2005. The gender similarities hypothesis. *American Psychologist* 60(6):581–92.

4. Vullioud, C., et al. 2019. Social support drives female dominance in the spotted hyaena. *Nature Ecology & Evolution* 3(1):71–76.

5. Tay, L., R. Su, and J. Rounds. 2011. People—things and data—

ideas: Bipolar dimensions? *Journal of Counseling Psychology* 58(3):424–40.

6. Hyde, The gender similarities hypothesis.

7. Fine, C. 2017. *Testosterone Rex: Myths of Sex, Science, and Society* New York: W. W. Norton & Company, 116.

8. Hyde, The gender similarities hypothesis.

9. Peters, M., et al. 2007. The effects of sex, sexual orientation, and digit ratio (2D:4D) on mental rotation performance. *Archives of Sexual Behavior* 36:251–60.

10. Johnson, P. A., S. E. Widnall, and F. F. Benya, eds. 2018. *Sexual Harassment of Women: Climate, Culture, and Consequences in Academic Sciences, Engineering, and Medicine.* A Consensus Study Report. Washington, DC: The National Academies Press.

11. Benner, K. June 30, 2017. "Women in Tech Speak Frankly on Culture of Harassment." *New York Times.* www.nytimes.com/ 2017/06/30/technology/women-entrepreneurs-speak-out-sexual-harassment.html. Accessed January 13, 2019.

Chapter 18. Gender Blending

1. Dongning, R., Z. Haotian, and F. Xiaolan. 2009. A deeper look at gender difference in multitasking: Gender-specific mechanism of cognitive control. 2009 Fifth International Conference on Natural Computation. https://ieeexplore.ieee.org/document/ 5364739. Accessed January 26, 2019; Mäntylä, T. 2013. Gender differences in multitasking reflect spatial ability. *Psychological Science* 24(4):514–20; Buser, T., and N. Peter. 2012. Multitasking. *Experimental Economics* 15(4):641–55.

2. Sanger, T. 2008. Queer(y)ing Gender and Sexuality: Transpeople's Lived Experiences and Intimate Partnerships. In Moon, L., ed. *Feeling Queer or Queer Feelings? Radical Approaches to Counseling Sex, Sexualities and Genders,* 72–88. London: Routledge.

3. Joel, D., et al. 2013. Queering gender: Studying gender identity in the normative population. *Psychology & Sexuality* 5(4):291–321.

4. Jacobson, R., and D. Joel. 2018. An exploration of the relations

between self-reported gender identity and sexual orientation in an online sample of cisgender individuals. *Archives of Sexual Behavior* 47:2407–26.

5. Statement from May 26, 2010, on de-psychopathologisation of gender variance. www.wpath.org/policies. Accessed January 15, 2019.

6. Jacobson, R., and D. Joel. 2018. Self-reported gender identity and sexuality in an online sample of cisgender, transgender and gender-diverse individuals: An exploratory study. *Journal of Sex Research* 56:249–63.

7. Siebler, K. 2012. Transgender transitions: Sex/gender binaries in the digital age. *Journal of Gay & Lesbian Mental Health* 16(1):74–99.

8. Martin, C. L., et al. 2017. A dual identity approach for conceptualizing and measuring children's gender identity. *Child Development* 88(1):167–82.

Chapter 19. Gender-Free Education

1. Martin, C. L., D. N. Ruble, and J. Szkrybalo. 2002. Cognitive theories of early gender development. *Psychological Bulletin* 128(6):903–33.

2. Shutts, K., et al. 2017. Early preschool environments and gender: Effects of gender pedagogy in Sweden. *Journal of Experimental Child Psychology* 162:1–17.

3. Barry, E. March 24, 2018. "In Sweden's Preschools, Boys Learn to Dance and Girls Learn to Yell." *New York Times.* www.nytimes.com/2018/03/24/world/europe/sweden-gender-neutral-preschools.html. Accessed January 13, 2019.

4. Ibid.

5. Shutts, op. cit.

6. *No More Boys and Girls: Can Our Kids Go Gender Free?* www.youtube.com/watch?v=wN5R2LWhTrY and www.youtube.com/watch?v=cp9Z26YgIrA. Accessed March 30, 2019.

7. National Association for Single-Sex Public Education. www.singlesexschools.org. Accessed March 30, 2019.

Chapter 20. Ungendering Our Children

1. Cabrera, N. J., J. D. Shannon, and C. Tamis-LeMonda. 2007. Fathers' influence on their children's cognitive and emotional development: From toddlers to pre-K. *Applied Development Science* 11(4):208–13.

2. Cools, S., J. H. Fiva, and L. J. Kirkebøen. 2015. Causal effects of paternity leave on children and parents. *The Scandinavian Journal of Economics* 117(3):801–28.

3. References to specific studies can be found in Paternity and parental leave policies across the European Union, a European Commission document. https://eur01.safelinks.protection.outlook.com/?url=https%3A%2F%2Fpublications.europa.eu%2Fen%2Fpublication-detail%2F-%2Fpublication%2Fa846 4ad8-9abf-11e8-a408-01aa75ed71a1%2Flanguage-en&data=0 2%7C01%7CLouise.McKeever%40octopusbooks.co.uk%7Cf4 db94fec89248a9b67c08d6c98c9831%7Cf881a2c50a8948318 1b1c7846c49594d%7C1%7C0%7C636918003034239775&s data=ND7vFWuod4zy9x8rBJVYYB203cF5Nbcox5giRCQP0 Ac%3D&reserved=0" https://publications.europa.eu/en/pu blication-detail/-/publication/a8464ad8-9abf-11e8-a408-01aa 75ed71a1/language-en); http://www.fatherhoodinstitute.org/2014/fi-research-summary-paternity-leave. Accessed 5 May 2019.

4. Arnalds, A. A., G. B. Eydal, and I. V. Gíslason. 2013. Equal rights to paid parental leave and caring fathers—the case of Iceland. *Icelandic Review of Politics and Administration* 9(2):323–44.

5. Dex, S., and H. Ward. 2007. Parental care and employment in early childhood. London: Equal Opportunities Commission, Working Paper 7; Huerta, M., et al. 2013. Fathers' leave, fathers' involvement and child development: are they related? Evidence from Four OECD Countries, OECD Social, Employment, and Migration Working Papers, 140.

Chapter 21. Gender Awareness

1. Joel, D., and D. Yarimi. 2014. Consciousness-raising in a gender conflict group. *International Journal of Group Psychotherapy* 64:48–69.
2. Survey of US adults conducted August 8–21 and September 14–28, 2017. "On Gender Differences, No Consensus on Nature vs. Nurture." www.pewsocialtrends.org/2017/12/05/on-gender-differences-no-consensus-on-nature-vs-nurture/. Accessed January 18, 2019.
3. Heilman, M. E., et al. 2004. Penalties for success: Reactions to women who succeed at male gender-typed tasks. *Journal of Applied Psychology* 89(3):416–27.

Chapter 22. Taking Action

1. Goldin, C., and C. Rouse. 2000. Orchestrating impartiality: The impact of "blind" auditions on female musicians. *American Economic Review* 90(4):715–41.
2. "Orchestrating Impartiality: The Impact of 'Blind' Auditions on Female Musicians." http://gap.hks.harvard.edu/orchestrating-impartiality-impact-%E2%80%9Cblind%E2%80%9D-auditions-female-musicians. Accessed March 20, 2019.
3. We thank Diana Lipton for this comment.
4. Vikhanski, L. 2016. *Immunity: How Elie Metchnikoff Changed the Course of Modern Medicine.* Chicago: Chicago Review Press.
5. Bigler, R. S., and C. Leaper. 2015. Gendered language: Psychological principles, evolving practices, and inclusive policies. *Policy Insights from the Behavioral and Brain Sciences* 2(1):187–94.
6. National Science Foundation, data for 2004–2014. www.nsf.gov/statistics/2017/nsf17310/static/data/tab5-1.pdf. Accessed January 13, 2019.
7. Klawe, M. October 6, 2015. "Opinion: How Can We Encourage More Women to Study Computer Science?" *Newsweek.* www.newsweek.com/how-can-we-encourage-more-women-study-computer-science-341652. Accessed January 13, 2019.

8. Editorial. 2016. Women need to be seen and heard at conferences. *Nature* 538(7625):290.

9. BiasWatchNeuro, "Summarizing 3.33 Years of BWN: We've Moved the Needle!" December 29, 2018. https://biaswatch-neuro.com/2018/12/29/summarizing-3-33-years-of-bwn-weve-moved-the-needle/. Accessed January 22, 2019.

10. Acosta, R. V., and L. J. Carpenter. 2014. *Women in Intercollegiate Sport: A Longitudinal, National Study Thirty-Seven-Year Update 1977–2014.* www.acostacarpenter.org. Accessed January 15, 2019.

11. "Women on Boards of Government and Public Companies," submitted to Aliza Lavie, chair of the Knesset's Committee on the Status of Women and Gender Equality (Hebrew). July 25, 2013. https://fs.knesset.gov.il/globaldocs/MMM/e2ef6d8d-f1f7-e411-80c8-00155d01107c/2_e2ef6d8d-f1f7-e411-80c8-00155d01107c_11_9948.pdf. Accessed January 22, 2019.

12. Aydin, R. September 30, 2018. "California Will Require Women on Corporate Boards under Bill Signed by Brown." *San Francisco Chronicle.* www.sfchronicle.com/business/article/California-will-require-women-on-corporate-boards-13270213.php. Accessed January 17, 2019.

Chapter 23. Vision

1. To appreciate the variability in the composition of human families, see *What the World Eats,* by Faith D'Aluisio, author, and Peter Menzel, photographer (Tricycle Press, 2008), featuring portraits of 25 families from 21 countries from all over the world. To learn about a family structure that's strikingly different from the Western heteronormative one, see "Where Women Reign: An Intimate Look Inside a Rare Kingdom," by Alexandra Genova, in *National Geographic.* www.nationalgeographic.com/photography/proof/2017/08/portraits-of-chinese-Mosuo-matriarchs/. Accessed January 19, 2019.

Suggested Reading

Eliot, Lise. 2009. *Pink Brain, Blue Brain: How Small Differences Grow into Troublesome Gaps—and What We Can Do About It*. Boston: Houghton Mifflin Harcourt.

Fausto-Sterling, Anne. 2000. *Sexing the Body: Gender Politics and the Construction of Sexuality*. New York: Basic Books.

———. 2012. *Sex/Gender: Biology in a Social World*. The Routledge Series Integrating Science and Culture. New York: Routledge.

Fine, Cordelia. 2010. *Delusions of Gender: How Our Minds, Society, and Neurosexism Create Difference*. New York: W. W. Norton & Company.

———. 2017. *Testosterone Rex: Myths of Sex, Science, and Society*. New York: W. W. Norton & Company.

Jordan-Young, Rebecca M. 2011. *Brain Storm: The Flaws in the Science of Sex Differences*. Cambridge, MA: Harvard University Press.

Richardson, Sarah S. 2013. *Sex Itself: The Search for Male and Female in the Human Genome*. Chicago: University of Chicago Press.

Rippon, Gina. 2019. *The Gendered Brain: The New Neuroscience That Shatters the Myth of the Female Brain*. London: Penguin Books.

Saini, Angela. 2018. *Inferior: How Science Got Women Wrong and the New Research That's Rewriting the Story*. Boston: Beacon Press.

Schiebinger, Londa. 1989. *The Mind Has No Sex? Women in the Origins of Modern Science*. Cambridge, MA: Harvard University Press.

Daphna Joel's Publications on Sex, Brain, and Gender

Joel, D., and R. Tarrasch. 2010. The risk of a wrong conclusion—on testosterone and gender differences in risk aversion and career choices. *Proceedings of the National Academy of Sciences, USA* 107:E19.

Joel, D. 2011. Male or female? Brains are intersex. *Frontiers in Integrative Neuroscience* 5:57.

Joel, D. 2012. Genetic-gonadal-genitals sex (3G-sex) and the misconception of brain and gender, or, why 3G-males and 3G-females have intersex brain and intersex gender. *Biology of Sex Differences* 3:27.

Joel, D., R. Tarrasch, Z. Berman, M. Mukamel, and E. Ziv. 2013. Queering gender: Studying gender identity in the normative population. *Psychology & Sexuality* 5:291–321.

Joel, D., and D. Yarimi. 2014. Consciousness-raising in a gender conflict group. *International Journal of Group Psychotherapy* 64:48–69.

Joel, D. 2014. Response to Nina K. Thomas and J. Scott Rutan: Is the personal political? And who benefits from believing it is not? *International Journal of Group Psychotherapy* 64:83–89.

Joel, D., and R. Yankelevitch-Yahav. 2014. Reconceptualizing sex, brain and psychopathology: Interaction, interaction, interaction. *British Journal of Pharmacology* 171:4620–35.

Joel, D. 2014. Sex, Gender, and Brain—A Problem of Conceptualization. In Schmitz, S., and G. Höppner, eds. *Gendered Neurocultures. Feminist and Queer Perspectives on Current Brain Discourses*, 169–86. Vienna: Zaglossus.

Joel, D., and R. Tarrasch. 2014. On the mis-presentation and misinterpretation of gender-related data: The case of Verma's human connectome study. *Proceedings of the National Academy of Sciences, USA* 111:E637.

Fine, C., D. Joel, R. Jordan-Young, A. Kaiser, and G. Rippon. December 15, 2014. "Why Males ≠ Corvettes, Females ≠ Volvos, and Scientific Criticism ≠ Ideology." Reaction to "Equal ≠ The Same: Sex Differences in the Human Brain." *Cerebrum*. http://dana.org/Cerebrum/Default.aspx?id=115816.

Joel, D., Z. Berman, I. Tavor, N. Wexler, O. Gaber, Y. Stein, N. Shefi, J. Pool, S. Urchs, D. Margulies, F. Liem, J. Hänggi, L. Jäncke, and Y. Assaf. 2015. Sex beyond the genitalia: The human brain mosaic. *Proceedings of the National Academy of Sciences, USA* 112(50):15468–73.

Joel, D. 2015. The NIH call to consider sex as a biological variable is conceptually "captured" in the "sex differences" paradigm. *Catalyst* 1:4–5.

Joel, D., A. Persico, J. Hänggi, J. Pool, and Z. Berman. 2016. Reply to Del Guidice et al., Chekroud et al., and Rosenblatt: Do brains of females and males belong to two distinct populations? *Proceedings of the National Academy of Sciences, USA* 113(14):E1969–70.

Joel, D., J. Hänggi, and J. Pool. 2016. Reply to Glezerman: Why differences between brains of females and brains of males do not "add up" to create two types of brains. *Proceedings of the National Academy of Sciences, USA* 113(14):E1972.

Joel, D., and A. Fausto-Sterling. 2016. Beyond sex differences: New approaches for thinking about variation in brain structure and function. *Philosophical Transactions of the Royal Society B* 371:20150451.

Joel, D. 2016. Captured in terminology: Sex, sex categories, and sex differences. *Feminism & Psychology* 26:335–45.

Joel, D., and M. M. McCarthy. 2017. Circumspective: Incorporating sex as a biological variable in neuropsychiatric research: Where are we now and where should we be? *Neuropsychopharmacology* 42:379–85.

Rippon, G., R. Jordan-Young, A. Kaiser, D. Joel, and C. Fine. 2017. Letter to the editor: Journal of neuroscience research policy on addressing sex as a biological variable: Comments, clarifications, and elaborations. *Journal of Neuroscience Research* 95:1357–59.

Fine, C., J. Dupre, and D. Joel. 2017. Sex-linked behavior: Evolution, stability, and variability. *Trends in Cognitive Sciences* 21:666–73.

Hyde, J. S., R. Bigler, D. Joel, C. Tate, and S. van Anders. 2018. The future of sex and gender in psychology: Five challenges to the gender binary. *American Psychologist* 74:171–93.

Jacobson, R., and D. Joel. 2018. An exploration of the relations between self-reported gender identity and sexual orientation in an online sample of cisgender individuals. *Archives of Sexual Behavior* 47:2407–26.

Jacobson, R., and D. Joel. 2018. Self-reported gender identity and sexuality in an online sample of cisgender, transgender and gender-diverse individuals: An exploratory study. *Journal of Sex Research* 56:249–63.

Joel, D., A. Persico, M. Salhov, Z. Berman, S. Oligschläger, I. Meilijson, and A. Averbuch. 2018. Analysis of human brain structure reveals that the brain "types" typical of males are also typical of females, and vice versa. *Frontiers in Human Neuroscience* 12:399.

Fine, C., D. Joel, and G. Rippon. 2019. Eight things you need to know about sex, gender, brains, and behavior: A guide for academics, journalists, parents, gender diversity advocates, social justice warriors, tweeters, Facebookers, and everyone else not otherwise specified. *Scholar & Feminist Online* 15.2.

Joel, D. In press. Beyond sex differences and a male-female continuum: Mosaic brains in a multidimensional space. In Lanzenberger, R., G. S. Kranz, and I. Savic, eds. *Handbook of Clinical Neurology,* 3rd Series, Sex Differences in Neurology and Psychiatry, chapter 2. Amsterdam: Elsevier.

Joel, D., A. Garcia-Falgueras, and D. Swaab. In press. The complex relationship between sex and the brain. *Neuroscientist.*

Index

Note: Page numbers in *italics* refer to illustrations.